SpringerBriefs in Computer Science

W0079444

SpringerBriefs present concise summaries of cutting-edge research and practical applications across a wide spectrum of fields. Featuring compact volumes of 50 to 125 pages, the series covers a range of content from professional to academic.

Typical topics might include:

- A timely report of state-of-the art analytical techniques
- A bridge between new research results, as published in journal articles, and a contextual literature review
- A snapshot of a hot or emerging topic
- An in-depth case study or clinical example
- A presentation of core concepts that students must understand in order to make independent contributions

Briefs allow authors to present their ideas and readers to absorb them with minimal time investment. Briefs will be published as part of Springer's eBook collection, with millions of users worldwide. In addition, Briefs will be available for individual print and electronic purchase. Briefs are characterized by fast, global electronic dissemination, standard publishing contracts, easy-to-use manuscript preparation and formatting guidelines, and expedited production schedules. We aim for publication 8–12 weeks after acceptance. Both solicited and unsolicited manuscripts are considered for publication in this series.

More information about this series at http://www.springer.com/series/10028

Yuan Yao • Xing Su • Hanghang Tong

Mobile Data Mining

 Springer

Yuan Yao
State Key Laboratory for Novel Software
Nanjing University
Nanjing, China

Xing Su
Graduate Center
City University of New York
New York, NY, USA

Hanghang Tong
Arizona State University
Tempe, AZ, USA

ISSN 2191-5768 ISSN 2191-5776 (electronic)
SpringerBriefs in Computer Science
ISBN 978-3-030-02100-9 ISBN 978-3-030-02101-6 (eBook)
https://doi.org/10.1007/978-3-030-02101-6

Library of Congress Control Number: 2018958912

This Springer imprint is published by the registered company Springer Nature Switzerland AG.
The registered company address is: Gewerbestrasse 11, 6330 Cham, Switzerland

Preface

We have witnessed a fast-moving technological revolution due to the emergence of powerful smartphones. Smartphones are no longer limited to a texting-calling device, but a personal intelligent assistant with increasing abilities in sensing, computing, and networking. To make full use of this intelligent assistant, various data are collected and analyzed to provide better services for the end users.

In this book, we introduce the essential steps for mobile data mining tasks, including data collection (Chap. 2), feature engineering (Chap. 3), and learning models (Chaps. 4–6). We also discuss some key challenges and possible solutions during the introduction of each step. Overall, this book can serve as a primer for beginners to gain a big picture of mobile data mining. It also covers some useful information for further in-depth research on the topic.

Nanjing, China   Yuan Yao
New York, NY, USA   Xing Su
Tempe, AZ, USA   Hanghang Tong
Aug 28, 2018

Contents

Acronyms

ARA	Average Resultant Acceleration
DFT	Discrete Fourier Transform
ETL	Extraction, Transformation, and Loading
FFT	Fast Fourier Transform
GMS	Global System for Mobile Communications
GPS	Global Positioning System
KMM	Kernel Mean Matching
Pegasos	Primal Estimated sub-GrAdient SOlver
SGD	Stochastic Gradient Descent
SMA	Signal-Magnitude Area
SVM	Support Vector Machine

Chapter 1
Introduction

Abstract Smartphones are usually equipped with various sensors by which the personal data of the users can be collected. To make full use of the smartphone data, mobile data mining aims to discover useful knowledge from the collected data in order to provide better services for the users. In this chapter, we introduce some background information about mobile data mining, including what data can be collected by smartphones, what applications can be built upon the collected data, what are the key steps for a typical mobile data mining task, and what are the key characteristics and challenges of mobile data mining.

1.1 Background

Nowadays, mobile devices especially smartphones are acting as intelligent assistants of their users by collecting various personal data and providing further services based on these data. The key advantage of this assistant lies in the fact that smartphone is an integral, unobtrusive part of our wearable devices, while it is usually equipped with multiple powerful sensors including GPS, accelerometer, gravity sensor, barometer, light sensor, gyroscope, and compass. Various types of data can be collected from these sensors (see Sect. 2.1 for more details). For example, GPS identifies the position of your smartphone, accelerometer records the accelerations along axes, and gyroscope measures the speed and angle of rotations. In a general sense, mobile data include not only sensor data but any data generated from mobile phones such as call log, message log, application usage data, etc. In this book, we mainly focus on the sensor data.

The mobile data generated by the sensors have already enabled some interesting and well-known applications. For example, based on the accelerations from accelerometer, we can compute the speed and direction that your phone is moving, and this is the reason why your smartphone can track your steps; likewise, the gyroscope functions when you are tilting the screen to steer in a racing game. In fact, the massive volume and convenient access of various mobile senor data have greatly facilitated a variety of more advanced mobile data mining applications, ranging from health and fitness monitoring, assistive technology and elderly care,

Y. Yao et al., *Mobile Data Mining*, SpringerBriefs in Computer Science,
https://doi.org/10.1007/978-3-030-02101-6_1

1

indoor localization and navigation, to urban computing and smart transportation, etc. The key idea behind these applications is to learn the facts of activities, events, and situations where the users are involved together with smartphones.

Existing literature has also extensively studied the mobile data mining topic, which can be divided into the following categories [43]: location-aware mining, context-aware mining, and social mining. Location-awareness is achieved by the connectivity or interaction with the ambient environment, and this category includes outdoor localization and indoor localization. The location information for outdoor localization is obtained through GPS, cellular tower connection, etc; in contrast, due to the constraint of no GPS and weak cellular reception, indoor localization is a much more active research area. In this area, various sources of data are utilized such as the wifi connections (along with the signal strength) [13, 23, 28, 34, 39, 61], the bluetooth connectivities [65], the acoustic sensor [22, 38, 51], the barometer [66], the magnetic sensor or compass [62], etc. Context-awareness means to infer the users' activities from smartphone data. Typical activities include walking on stairs, sitting, sleeping, driving, and jogging. Usually, detected activities are further used for other tasks in both individual level and aggregation level. In the individual level, the detected walking speed may reveal the mood of a person [37], the accelerations in driving may reveal the person's driving style [16], the distribution of activities may indicate the user's living style [31]; in the aggregation level, the detected activities of a set of persons can be used in traffic characterization and control [12, 20], carbon emission estimation [40], etc. The third social mining research involves the social media (e.g., the location based twitter data [3]) and focuses on the aggregation of users' mobilities and activities. Examples in this category includes the modeling of the interplay between human's mobility and social ties [25, 54, 59].

1.2 Typical Applications

In this section, we discuss some typical mobile data mining applications. As mentioned above, a smartphone can provide sensor data which can be easily processed to further obtain knowledge concerning the motion of the device and the ambient environment. Based on these data, mobile data mining solutions can be applied either as the core method or an assistive technology in many applications. For example, learning the pattern of daily smartphone locations via cellular tower connection records in a city could help estimate the transportation volume and predict the traffic of the city. Smartphone based activity recognition could help the public health agency on outdoor activity survey. Based on the fields towards which mobile data mining is applied, the main applications fall into the following four categories.

- **Urban computing.** Urban computing is an emerging field, which uses the data that are generated in cities from different sources such as traffic flow, human mobility, and geographical data to help the modernization of people's lives and tackle the urban issues such as traffic congestion, energy consumption, etc. Smartphones play an important role in urban computing. On one hand, the interaction between humans and the urban environment is reflected by the smartphone and its connection to cellular towers and environment wifis. The analysis of these data could help urban planning such as gleaning the underlying problems in transportation networks [70], discovering the functional regions and the city boundary [47, 68], etc. One the other hand, smartphone data are used to identify users' transportation modes. This leads to applications in urban transportation survey [40], daily traffic monitor [67], public transportation system design [35, 57], etc. A full survey of urban computing can be found in [71].

- **Healthcare.** Mobile mining solution is also used in healthcare related fields. In the individual level, the activity recognition and travel mode identification using smartphones could reveal the user's living style such as the working hours of the user, the active status of the user, the main transportation tools s/he is using, etc. The activity recognition can also help in elderly care such as fall detection [26] and rehabilitation assistant. More and more health related smartphone apps are developed and launched with latest technologies that assist users in monitoring their sleep quality [1], logging the walking steps [2], etc. In the aggregation level, the mobile mining techniques provide health related research useful information such as public physical activities [50].
- **Location-based search and advertising.** Mobile mining is heavily used in location based services such as maps, navigation, and social search (e.g. waze, yelp, and foursquare). Location can also assist more accurate marketing and advertising [18, 60]. For example, digital advertisement uses mobile mining techniques to target audiences by learning the patterns of users' behaviors such as transportation behavior and social behavior [17, 21, 36, 45, 46].
- **Privacy and security.** Unfortunately, mobile data mining is also used in malicious ways such as probing users' privacy, or stealing information. For example, through the vast data of some smartphone usage, one can accurately infer the activities or locations of users [19, 41]. Some mobile malware uses mobile mining technology to "learn" the user's keyboard input [64]. This poses challenges in technology related legislation.

In this book, we mainly focus on the application of urban computing. To be specific, when presenting the ideas and results, we focus on the scenarios of identifying users' travel modes in urban transportation with smartphone sensors. Meanwhile, the rationale behind is applicable to the other scenarios and applications.

1.3 Steps, Characteristics, and Challenges

As shown in Fig. 1.1, there are three key steps for developing a typical mobile data mining application, i.e., (1) data capturing and preprocessing, (2) feature engineering, and (3) modeling. The applications sometimes provide feedback to the data capturing step so that the model can be updated.

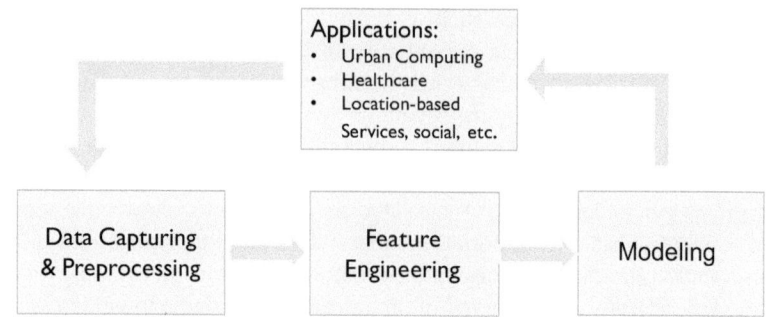

Fig. 1.1 Typical steps of mobile data mining

The data capturing step aims to collect and transform the raw mobile data to its ready-to-use form. This step is also known as ETL (i.e., extraction, transformation, and loadling). The raw mobile data are extracted from various sources, transformed into a clean format, and then loaded into the warehouse where the mining activities take place. For the feature engineering step, the key is to design the "right" features for specific tasks. After the features are extracted, the modeling step uses machine learning models such as classifiers or regressors to make the predictions or decisions.

Since the data are collected via smartphones and the algorithms or methods are executed on the smartphones as well, mobile data mining needs to pay special attention to the balance of factors such as accuracy, network stability, bandwidth availability, storage capacity, computational time, response time, and battery capacity. For example, continuously sensing may achieve higher accuracy while it would quickly drain the battery.

Stemming from the three key steps in Fig. 1.1, we summarize the following key challenges related to (C1) data capturing and preprocessing, (C2) feature engineering, and (C3) model training and adaptation.

In data capturing and preprocessing, since different activities tend to have different characteristics of motion patterns (e.g., speed variations, acceleration rates, and rotations) and even the environmental patterns (e.g., magnetic field change and ambient air pressure), it is important to select the most suitable sensors for specific tasks. For example, the accelerometer sensor, as the exclusive data source in most existing smartphone-based activity recognition research, is inadequate to differentiate some travel modes such as driving and taking subway. Moreover, the collected data often accompany with various noises which might overweigh the discriminative power.

In feature engineering, the feature extraction needs to balance a number of potentially conflicting factors such as the recognition accuracy, the computational time, the response time, and the battery consumption. For example, in order to segment the raw sensor readings (which are essentially time series data), we need to decide the appropriate sampling rate and segmentation length. A longer segment might lead to more powerful features (e.g., the discriminative FFT coefficients), yet it also requires more computational time to extract such features and provide online responses. In general, it is desirable to examine all the available sensors to have more discriminative features. However, it inevitably comes with a cost in terms of the battery consumption.

In modeling, the main challenges are model adaption and battery consumption. The model adaption concerns the following two major questions: (1) how to adapt a model which might be trained on a generic population to a specific user, and (2) how to update a model efficiently over time to accommodate newly collected labeled data. When answering these two questions, we need to always keep in mind the challenges of battery consumption. Consider the case of detecting whether the user is driving or not. Some drives may drive cautiously, while others might drive more aggressively with frequent and sudden accelerations and decelerations. Therefore, the detection model should be sufficiently adaptive to different users, especially under the constraints that the labeled data for each user are not sufficient in many real scenarios. Additionally, when the valuable labeled data are available in an online fashion, it is more appealing to update the model accordingly instead of retraining the model.

1.4 Roadmap

In response to the above challenges of mobile data mining, the organization of this book is summarized in Fig. 1.2.

Chapter 2 introduces the data capturing and data processing. For data capturing, we illustrate how to collect the data via a smartphone App. For data processing, we first analyze the typical patterns from the collected data, and then provide a set of comprehensive denoising methods. The output of this step are the time series data of different sensors' readings.

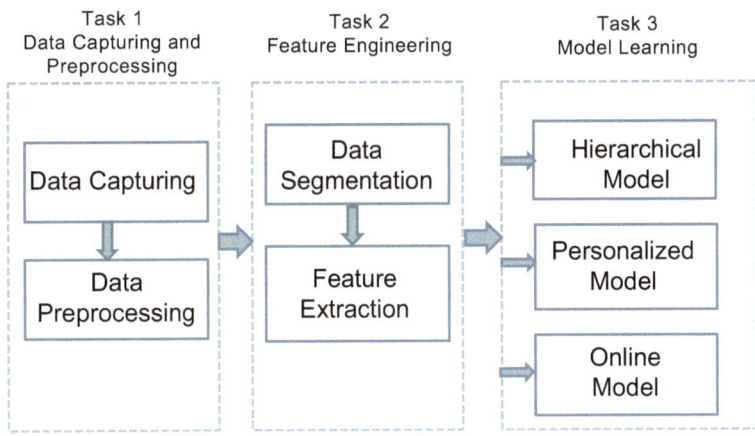

Fig. 1.2 The roadmap

Chapter 3 describes the data segmentation and feature extraction methods that carefully balance between the prediction accuracy, the online response time, and the battery consumption. We first slice the time series data into data segments and then extract the most discriminative features from these segments. Meanwhile, special considerations should be placed on the balance of effectiveness and efficiency.

After the features are properly extracted, Chaps. 4–6 present three learning models to accommodate different situations to save energy consumption, to provide personalized treatment, and to update the model in an online fashion, respectively. In particular, Chap. 4 realizes that not all mobile data mining tasks require all the features, based on which it introduces a hierarchical learning model to save energy consumption. Chapter 5 tackles the personalization challenge by appropriately "borrowing" the data from users whose data are similar to current user. Chapter 6 describes the online learning model to update the learned model when a small portion of new labeled data arrive. In these three chapters, we take the travel mode detection scenario as an example with the goal of recognizing the following travel modes: *driving a car, walking, jogging, bicycling, taking a bus*, and *taking a subway*.

Chapter 2
Data Capturing and Processing

Abstract The first step towards mobile data mining is collecting smartphone sensors' readings, and processing the raw data in order to remove various noises. The key questions that need to be answered during data capturing and processing include: what data should be collected and how to collect such data, and what are the characteristic patterns behind these data and how to clean the data so that these patterns can be easily identified. In this chapter, we first introduce the smartphone sensors and their readings. Next, to facilitate discussions, we present the details of data collection and data denoising with an example of travel mode detection.

2.1 Smartphone Sensors

Data capturing is the process of using the smartphones to record data. These data can be the smartphone sensors' readings, the activity logs by certain smartphone apps that record how the user is interacting with his/her phone, etc. In this book, we mainly focus on the smartphone sensors' readings. The smartphone sensors can be classified into four categories [4]: motion sensors, environmental sensors, position sensors, and connection sensors.

- **Motion Sensors.** Motion sensors measure acceleration forces and rotational forces along three axes of the phone's coordinates. This category includes accelerometers, gravity sensors, gyroscopes, and rotational vector sensors.
- **Environmental Sensors.** Sensors in this category measure various environmental parameters, such as ambient air temperature and pressure, illumination, and humidity. This category includes barometers, photometers, and thermometers.
- **Position Sensors.** Position sensors measure the physical position of a device. This category includes orientation sensors and magnetometers.
- **Connection Sensors.** Connection sensors provide the solutions for smartphones to connect and interact with other devices via various protocols. This category includes Bluetooth, GPS sensors, wireless sensors, standard cellular connection modulars.

Table 2.1 A summary of smartphone sensors

Sensor name	Data collected	Dimensions	Unit
Accelerometer	Acceleration	x, y, z	g-force
Gravity sensor	Gravity	x, y, z	m/s^2
Gyroscope	Rotation rate	$x, y, z,$ x_calibration, y_calibration, z_calibration	rad/s
Magnetometer	Magnetic field	$x, y, z,$ x_calibration, y_calibration, z_calibration	μT
Barometer	Ambient air pressure	1	hPa
Rotation Sensor	Rotation degree	Azimuch: rotation around z axis	Degree
	(y axis pointing to magnetic	Pitch: rotation around x axis	
	north as the default)	Roll: rotation around y axis	
Proximity sensor	Relative distance from an object to the view screen of a device	1	cm
Light sensor	The ambient light level	1	lx
Humidity sensor	The relative ambient humidity	1	Percentage
Temperature sensor	Ambient temperature	1	Celsius
GPS sensor	Geographical description of current location and estimated speed	Latitude, longitude, speed	Degree, m/s

A summary of the smartphone sensors and their sensing data are shown in Table 2.1. The data range and resolution are different from phone to phone. Some sensors may have very high resolution and cost more, while some others return an estimated result indicating the level (e.g., some proximity sensors return a binary result indicating far or near to the user's face).

2.2 Data Collection

In data capturing, the main question to answer is how to collect data and what to collect. The answer to this question depends on the specific mobile data mining task. For example, the accelerometer data and GPS/GSM data are the main data sources in most of the existing smartphone-based travel mode detection systems [29]. However, such data are not only sensitive as they reveal the users' locations, but also unstable and even inapplicable in certain scenarios such as the underground transportation. Moreover, the accelerometer alone is often inadequate to effectively distinguish all travel modes. For example, in wheeled travel modes (e.g., buses, cars, and subways), travelers are sitting in their seats for most of the time, resulting very similar accelerometer readings. Therefore, the data from multiple sensors usually need to be collected for a given task.

In practice, the data are collected by mobile apps, and researchers start with developing their own apps with volunteers to record the sensors' data as well as the labels. In this section, we take travel mode detection as an example, and illustrate how to collect the suitable sensor data for this task.

To decide which sensors' data should be collected, we first need to understand the typical patterns from the data. The accelerometer data collected during jogging fluctuate more heavily than in other travel modes, and such data are the most frequently used sensor data in smartphone based activity recognition. The ambient air pressure detected by barometers largely depends on the space of the transportation tools, and other factors such as the air conditioner, door opening and closing, number of passengers, etc. Therefore, the air pressure detected during traveling with buses has a smaller mean value than that of traveling with cars. The magnetometer provides a good indicator of different patterns related to the environment's magnetic field and its changes during travel. For example, a subway system uses magnet to brake, and the magnetic field readings show some oscillatory patterns while traveling on subways. Also, there are more mobile devices on a bus than in a car which would also cause the change in the detected magnetic field. All these differences are revealed in the magnetometer readings and thus can be leveraged to identify the travel mode.

We summarize the collected data for travel mode detection as follows.

Accelerometer Accelerometer readings measure the changes in velocity along the x, y, and z axes of the cell phone, as is shown in Fig. 2.1. Accelerometer provides an important reference to detect the pattern of a user's body movement.

Gravity Sensor Gravity sensor readings return the gravity as measured along each axis of the cell phone. If the phone is put on the table with y axis facing the sky, the reading on y axis would be roughly $-9.8 \, \text{m/s}^2$ while the readings on the other 2 axes would be around $0.0 \, \text{m/s}^2$.

Barometer Barometer readings return the detected ambient air pressure. In [42], Muralidharan et al. conducted an experiment showing that the pressure detected by the smartphone barometer would change with the building structure and type, and such a pattern is able to be learned. It can also be used as a discriminative factor in other mobile data mining tasks such as travel mode detection.

Gyroscope The gyroscope measures the rate of rotation around the three axes. Figure 2.2 shows the standard direction of measures for rotation along x, y and z axes.

Light Sensor It measures the ambient light level in SI lux units. In Android phones this value can be directly obtained [5]. In iPhone it is discouraged to use [10];

Fig. 2.1 Acceleration
measure of iPhone [8]

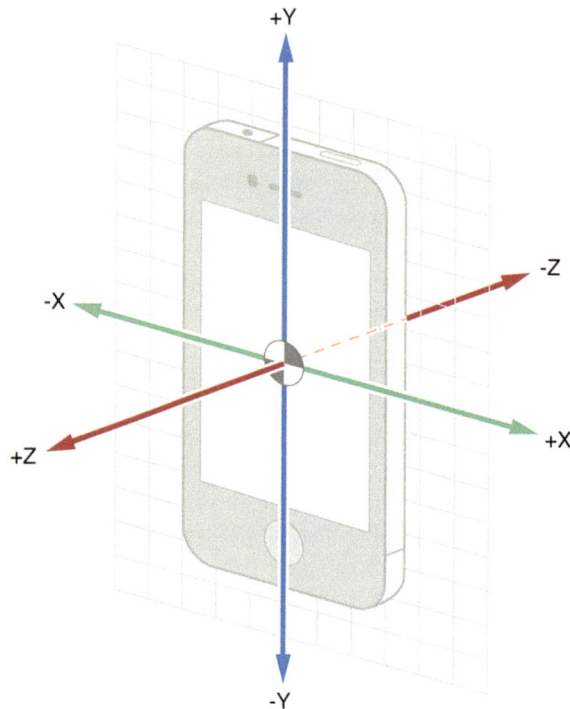

Fig. 2.2 Rotation measure of
iPhone [8]

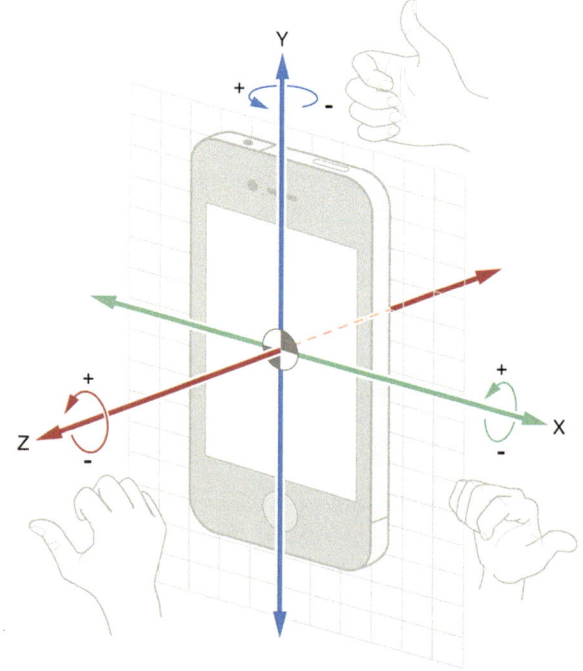

therefore, we use screen brightness reading which is the brightness level of the screen instead [9]. This reading is adjusted by the ambient light level when the phone is unlocked and thus it is similar to the light sensor.

Magnetometer Magnetometer measures the earth's geomagnetic field. The smartphones provide both raw readings of the magnetic field as well as the calibrated readings. The calibrated magnetic field usually filters out the bias introduced by the device and, in some cases, its surrounding fields [6, 7]. The magnetometer also returns the hard iron bias values separately for customized calibration.

In all the experimental results shown in this book, we use the data collected in the past few years. Specially, we collect the data three times. The first time was in the July of 2014, the second time data collection happened in December 2014, and the last one happened in July–August 2016. In total, we have 16 volunteers collecting data in Buffalo, NY, New York City (NYC), NY, and College Station (CS), TX, while traveling with six major travel modes. The collected data have variation in user's gender, travel city, travel season, as well as the phone models. The sensors and their sensing settings are shown in Table 2.2 and the data details for each mode are shown in Table 2.3.

Table 2.2 Sensors' setting

Sensor used	Data name	Freq.	Dimensions
Accelerometer	Acceleration (m/s^2)	16 Hz	x, y, z
Gravity sensor	Gravity (m/s^2)	16 Hz	x, y, z
Gyroscope	Rotation rate (rad/s)	16 Hz	$x, y, z,$ $x_{calib},$ $y_{calib},$ z_{calib}
Magnetometer	Magnetic field (μT)	1 Hz	x, y, z $x_{calib},$ $y_{calib},$ z_{calib}
Barometer	Ambient air pressure (hPa)	1 Hz	1
Light sensor	Ambient light level (lx)	1 Hz	1

Table 2.3 Data collection details

Mode	#Samples	Duration(min.)	Gender dist.	City dist.
Bike	20,606	72.1	1 female, 2 males	NYC, Buffalo.
Bus	44,080	147	1 female, 3 males	NYC, Buffalo
Car	227,967	760	6 females, 6 males	NYC, Buffalo, CS
Jog	16,838	56.2	2 females, 5 males	NYC, Buffalo, CS
Subway	72,643	242.1	6 females, 3 males	NYC, Buffalo
Walk	95,701	319	7 females, 5 males	NYC, Buffalo, CS

2.3 Data Denoising

The raw data collected from sensors are often accompanied with various noises. Without being appropriately denoised, the negative impact of such noises might outweigh the additional discriminative power embedded in the sensors. Therefore, the main task in data preprocessing is to minimize the noises in the raw sensor readings. Here, we provide a comprehensive set of techniques to remove various noises in the raw data, as summarized in Table 2.4.

Denoising #1: Data Rotation Among all the sensors' readings we collected, acceleration is measured along the phone axes whose coordinates are determined by the phone's position and heading direction. Since it is unrealistic to coordinate all the travelers to have the same phone position and heading direction, we need to rotate the readings back to a standard coordinate system before any further calculation. Here, we define the standard coordinate system to be y axis perpendicular to the earth pointing toward the sky and z axis pointing to the magnetic north. The x axis is determined in a right-handed coordinate system. Magnetic field and gravity are used to rotate the readings from the phone's coordinates to the standard coordinates.

The readings of both calibrated and uncalibrated ambient magnetic field are accessible. According to [49], the differences between calibrated and uncalibrated magnetic field data are as follows. The hard iron calibration is reported separately in uncalibrated magnetic field, while calibrated reading includes the calibration in measurement. For example, the calibrated and uncalibrated magnetic field measures at the same time by one of our smartphones are shown in Table 2.5. The *TYPE_MAGNETIC_FIELD* indicates the ambient magnetic field measured in 3-dimensional space at time tick 633. The *TYPE_MAGNETIC_FIELD_UNCALIBRATED* indicates the hard iron calibrated magnetic field in three dimensions, as well as the bias that is calibrated along each dimension.

We denote the magnetic field vector by **M**, and the magnetic field magnitude is calculated as

$$\|\mathbf{M}\| = \sqrt{M_x{}^2 + M_y{}^2 + M_z{}^2} \tag{2.1}$$

Table 2.4 Denoising techniques and the targeted noises

De-noising techniques	Noises to remove
Data rotation	Noises caused by phones' different positions and heading directions
Winsorization	The outliers in sensor reading such as spikes in data, or noises caused by other sudden events (e.g. phone dropped to the floor)
Gaussian smoothing	The high frequency noise (e.g. white noise naturally exists in sensors)
Normalization	Difference in data scale of different data sources

Table 2.5 Magnetic field readings

Sensor field	Timestamp	Value name	Value(μT)
TYPE_MAGNETIC_FIELD_UNCALIBRATED	633	$x_{uncalib}$	100.7734
TYPE_MAGNETIC_FIELD_UNCALIBRATED	633	$y_{uncalib}$	−53.8651
TYPE_MAGNETIC_FIELD_UNCALIBRATED	633	$z_{uncalib}$	461.7340
TYPE_MAGNETIC_FIELD_UNCALIBRATED	633	x_{bias}	69.1016
TYPE_MAGNETIC_FIELD_UNCALIBRATED	633	y_{bias}	−26.0590
TYPE_MAGNETIC_FIELD_UNCALIBRATED	633	z_{bias}	497.7585
TYPE_MAGNETIC_FIELD	633	x_{calib}	31.6711
TYPE_MAGNETIC_FIELD	633	y_{calib}	−27.8061
TYPE_MAGNETIC_FIELD	633	z_{calib}	−36.0245

If we calculate the calibrated magnetic field using x_{calib}, y_{calib}, z_{calib}, the magnitude is 55.44 μT, which is close to the value 53.723 μT measured and recorded by Center NGD [15] in Buffalo on the day our data were collected. We use this fact for the first step of phone coordinates rotation.

Gravity reading is then used for the rotation. Gravity reading is recorded along three axes. For example, a sample of gravity reading in the format of $[t, x, y, z]$ is $[677, -0.850, 5.090, 8.339]$. We denote the gravity by $\mathbf{G}$, and the gravity magnitude is calculated by

$$\|\mathbf{G}\| = \sqrt{G_x^2 + G_y^2 + G_z^2} \tag{2.2}$$

The gravity magnitude of the sample reading is 9.807 m/s^2, which is close to the gravity 9.804 m/s^2 of Buffalo recorded in [44].

We denote the raw gravity and magnetic field reading by $\mathbf{G_0} = [G_{x0}, G_{y0}, G_{z0}]$ and $\mathbf{M_0} = [M_{x0}, M_{y0}, M_{z0}]$. In the standard coordinate system, the earth gravity is $\mathbf{G_s} = [G_{xs}, G_{ys}, G_{zs}]$ where $G_{xs} = 0$, $G_{ys} = \sqrt{G_{x0}^2 + G_{y0}^2 + G_{z0}^2}$, $G_{zs} = 0$, and magnetic field is $\mathbf{M_s} = [M_{xs}, M_{ys}, M_{zs}]$, where $M_{xs} = 0$, $M_{ys} = 0$, $M_{zs} = \sqrt{M_{x0}^2 + M_{y0}^2 + M_{z0}^2}$. $\mathbf{R_1}$ and $\mathbf{R_2}$ are the two rotation matrices we use. $\mathbf{R_1}$ rotates the raw gravity reading $\mathbf{G_0}$ to the Earth gravity $\mathbf{G_s}$, and $\mathbf{R_2}$ rotates the raw reading of magnetic field $\mathbf{M_0}$ to $\mathbf{M_s}$, i.e.,

$$\mathbf{R_1 G_0^T} = \mathbf{G_s^T}$$
$$\mathbf{R_2 M_0^T} = \mathbf{M_s^T} \tag{2.3}$$

Using the above equations, we can obtain $\mathbf{R_1}$ and $\mathbf{R_2}$. Now assume we have the raw reading of acceleration $\mathbf{A_0}$ at the same time. We use the following equation to rotate the acceleration into the standard coordination value $\mathbf{A_s}$.

$$\mathbf{R_2 R_1 A_0^T} = \mathbf{A_s^T} \tag{2.4}$$

By rotation, the sensor reading will be *position-independent*.

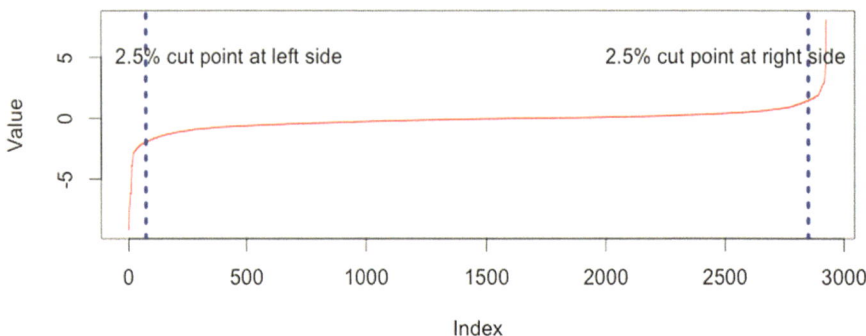

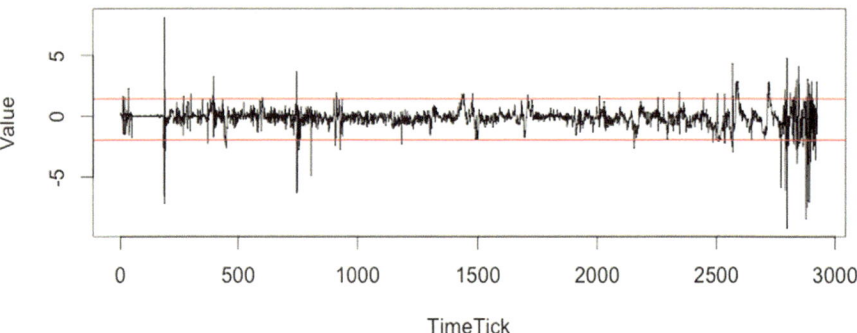

Fig. 2.3 Winsorization of acceleration data (5% along the *y* axis)

Denoising #2: Winsorization The winsorization is used to reduce the possible spurious outliers in the data. Figure 2.3 shows the distribution of acceleration data along *y* axis while traveling by a car. The top plot is the sorted acceleration in the ascending order. The dash lines indicate the 2.5% cutting point at both sides. The bottom plot is the unsorted raw acceleration along time. The two horizontal solid lines indicate the upper bound (97.5%) and the lower bound (2.5%) of the data values. We can see that in the top plot, the dash lines cut at the point where the data curve on the left side starts to change from very steep to very flat, and on the right side after the dash line it starts to change from very flat to very steep. Meanwhile, in the bottom plot, the horizontal lines keep the characteristics of the curve and only eliminate some outliers and noises. We have a similar observation in most of our data. Therefore, we recommend to chose 5% as the default fraction for winsorization.

Denoising #3: Gaussian Smoothing In winsorization, we eliminate the overall outliers within the data. To analyze the data value's change in a relatively longer time period, it is also important to eliminate the white noise and high frequency fluctuation. For example, if we want to learn the direction change during moving

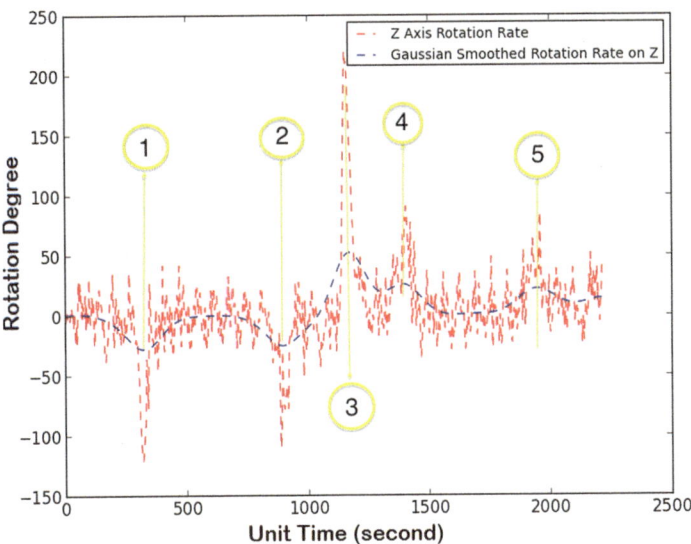

Fig. 2.4 Gaussian smoothed rotation rate

(e.g., when the car takes a turn, which direction it turns to), we need to analyze the readings from the rotation sensor (e.g., the red dash line in Fig. 2.4). Roughly, the readings contain three kinds of changes: the white noise in highest frequency, the fluctuation in second highest frequency which is caused by the phone's movement together with the carrier's body, and the trend change which includes the average value increasing/decreasing and the peak points. While analyzing the readings' trend change, we would like to eliminate the first two kinds of changes and keep the third one. The solution is to use Gaussian filter to smooth the data. It works as a low-pass filter and attenuates high frequency signal periods in the data. We use the segment length as the smooth parameter σ. The blue dash line in Fig. 2.4 includes the results after smoothing. We can see that the filtration drops all the high frequency oscillations most of which come from noises, and leaves only the main up and down trends. We compared the smoothed line with the recorded actual turning events. All the local extrema (1–5 points marked with a yellow line) in Fig. 2.4 are when the turning events actually happened.

Denoising #4: Normalization Finally, data collected by different sensors measuring different aspects of traveling events are quite diversified in both range and value. For example, the accelerometer data are between ±40 m/s^2, while the magnitude of magnetic field could reach 500 mT. In order to compare the data from different sensors, we normalize them between 0 and 1.

2.4 Summary

In this chapter, we first discuss the available sensors in smartphones and the corresponding data that can be collected from these sensors. After that, we take the travel mode detection problem as an example, and illustrate how to determine the possibly useful sensors for this task. We also provide a set of denoising techniques for preprocessing the collected data. Most of the discussed techniques are applicable to other data mining tasks.

Chapter 3
Feature Engineering

Abstract So far, the collected data are time series data of different sensors' readings. To make use of these time series in the following learning models, we usually need to first slice the time series data into data segments, and then extract features from these segments. Meanwhile, the data segmentation and feature extraction also affect the aspects like energy efficiency, model accuracy, and response time. In this chapter, we first discuss the data segmentation method, and then introduce the feature extraction which extracts features from a segment with the principle that the extracted features should be informative and discriminative. To further save time, we also discuss the feature selection method which selects a subset of the features for the current task.

3.1 Data Segmentation

In this book, we refer to the readings from all sensors at a single time as one "data sample". It is the unit data generated at one time tick, and a data segment contains one or more data samples. The key challenge in data segmentation is to find a proper way to slice the data providing the sensor sampling rate. If the length of a segment is too short, it might be less informative and result in a less effective model. On the other hand, increasing the length of the data segments means the system requires more time to obtain enough readings.

Figure 3.1 illustrates the commonly used slide-window segmentation, which leads to a fast response while preserving the information of each segment. To construct one segment, the slide-window mechanism extends the newly registered data into a segment by including a certain amount of the most recent cached data. As shown in the figure, if a segment consists of three data samples and is configured to take two historical data samples, at time t new data sample "$Sample_t$" is combined with the historical data "$Sample_{t-2}$" and "$Sample_{t-1}$". The segmentation window slides forward as the "$Sample_{t+1}$" comes in at time $t + 1$. "$Sample_{t-1}$" and "$Sample_t$" are combined with "$Sample_{t+1}$" as a new segment. The ratio of the new data and the cached data in a segment is configurable. The use of cached data

© The Author(s), under exclusive license to Springer Nature Switzerland AG 2018   17
Y. Yao et al., *Mobile Data Mining*, SpringerBriefs in Computer Science,
https://doi.org/10.1007/978-3-030-02101-6_3

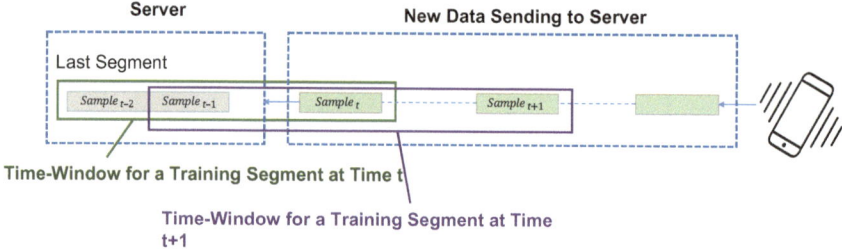

Fig. 3.1   Using slide window for data segmentation

will serve to avoid the boundary issue in the data. Without it, some discriminative patterns might be lost simply because a significant amount of consecutive readings would be divided into separate segments.

Based on the slide-window segmentation, there are two factors for a proper length of a data segment: the sampling rate and the cached data size. As for the sampling rate, it determines how long it takes to collect enough data for one segment and how much information one segment contains. For example, if smartphone A collects data with sampling rate 200 ms/reading and smartphone B collects data with sampling rate 100 ms/reading, a segment of length 10 contains information within 2 s of phone A while it contains information within 1 s of phone B. The cached data size, on the other side, determines how much new information is included in one segment. The larger the ratio of cached data, the shorter it takes to form a data segment. However, it also places more delay for event detection with larger ratio of cached data.

3.2   Feature Extraction

In feature extraction, the constraints are the computational cost and the battery consumption. This is especially true when the features are extracted from multimodality sensors and the sensing task for smartphones is battery draining. Following data segmentation, we can extract features in both time and frequency domains. Table 3.1 summarizes the mostly used features in mobile data mining literature.

Take the travel mode detection task as an example. This task involves the raw data including acceleration, magnetic field, and air pressure data. Acceleration and magnetic field data can be analyzed in both the time domain and the frequency domain. Pressure is analyzed in the time domain only since the main change in ambient air pressure largely depends on the space of the transportation tools and the statistical features such as max, min, standard deviation, mean would be enough to describe the characteristics. Then, based on the feature extraction methods in Table 3.1, we can extract various features for different mobile data mining tasks. Take the travel mode detection problem as an example, we can extract 161 features as summarized in Table 3.2.

Table 3.1 Feature extraction in mobile data mining

Category	Features	Description
Time domain features	Mean, max, min, std	The mean, max, minimum values and the standard deviation of the data segment
	Correlation	The correlation of data among each axis pair
	Signal-magnitude area	SMA is calculated as the sum of the magnitude of the three axes acceleration within the segment window [32]
	Average resultant acceleration	ARA is the average of the square root of the sum of the values of each axis
Frequency domain features	Energy	Calculated as the sum of the squared Discrete Fourier transform (DFT) component magnitudes. Ravi in [48] uses the normalized energy divided by the window length
	Entropy	Calculated as the normalized information entropy of the DFT components, and it helps in discriminating the activities with the similar energy features [11]
	Time between peak	The time between the peaks in the sinusoidal waves [33]
	Binned distribution	This feature is essentially the histogram of the FFT

3.3   Feature Analysis and Sensor Selection

For travel mode detection, while all the 161 extracted features in Table 3.2 seem to be effective in distinguishing different travel modes, it comes with a cost in terms of both the computation and battery consumption. Intuitively, not all the travel modes need all the features in learning. For example, the travel modes that involve more motion activities might rely on motion based features more than others in classification. In this section, we provide a feature selection method to analyze the feature importance by applying group lasso in the classification model to explore the possibilities of eliminating the sensor usage. That is, we put the features from the same sensor into one group, and select the groups that are potentially useful for the prediction task.

The lasso is a very popular technique for variable selection for high dimensional data. Group lasso [69] is a generalization of the lasso for doing group-wise variable selection. For a given dataset that contains the data samples (denoted by $\mathbf{X}$) and the corresponding label vectors (denoted by $\mathbf{y}$), group lasso aims to solve the penalized least squares,

$$\min_{\beta} \frac{1}{2} \|\mathbf{y} - \mathbf{X}\beta\|_2^2 + \lambda \sum_{l=1}^{L} \sqrt{p_k} \|\beta^{(k)}\|_2, \lambda > 0, \qquad (3.1)$$

Table 3.2 Extracted features for travel mode detection

Name	Description
$x_{max}, x_{min}, x_{std}, x_{avg}$ $y_{max}, y_{min}, y_{std}, y_{avg}$ $z_{max}, z_{min}, z_{std}, z_{avg}$	The statistical description of acceleration data segments along x, y and z axes, respectively
$x_{offset}, y_{offset}, z_{offset}$	The offset of the DFT results of the x, y and z axes of acceleration data segment, respectively
$x_{freq}, y_{freq}, z_{freq}$	The principle frequency of the acceleration along x, y and z axis, respectively
$xe_{std}, ye_{std}, ze_{std}$	The standard deviation of energy vector calculated from x, y and z axes acceleration of the data segments, respectively
$x_1, x_2, ..., x_{10}$ $y_1, y_2, ..., y_{10}$ $z_1, z_2, ..., z_{10}$	Histogram of normalized energy of the acceleration along x, y and z axes of the data segment, respectively
$xpk_1, xpk_2, xpk_3, xpk_4,$ $ypk_1, ypk_2, ypk_3, ypk_4,$ $zpk_1, zpk_2, zpk_3, zpk_4$	The time zone (indexes of time tick) of the top four data peaks of x, y and z axes acceleration, respectively
$a_{max}, a_{min}, a_{std}, a_{avg}$	The statistical description of total acceleration
$a_{offset}, a_{freq}, ae_{std}$	The offset,principle frequency and standard deviation of the DFT analysis on total acceleration
$a_1, a_2, ..., a_{10}$	Histogram of normalized energy of the total acceleration
$apk_1, apk_2, apk_3, apk_4$	The time zone (indexes of time tick) of the top four data peaks of total acceleration
$rx_{max}, rx_{min}, rx_{std}, rx_{avg}$ $ry_{max}, ry_{min}, ry_{std}, ry_{avg}$ $rz_{max}, rz_{min}, rz_{std}, rz_{avg}$	The statistical description of rotation data segments along x, y and z axes, respectively
$rx_{offset}, ry_{offset}, rz_{offset}$	The offset of the DFT results of the x, y and z axes of rotation data segment, respectively
$rx_{freq}, ry_{freq}, rz_{freq}$	The principle frequency of the rotation along x, y and z axis, respectively
$rxe_{std}, rye_{std}, rze_{std}$	The standard deviation of energy vector calculated from x, y and z axes rotation of the data segments, respectively
$rx_1, rx_2, ..., rx_{10}$ $ry_1, ry_2, ..., ry_{10}$ $rz_1, rz_2, ..., rz_{10}$	Histogram of normalized energy of the rotation along x, y and z axes of the data segment, respectively
$rxpk_1, rxpk_2, rxpk_3, rxpk_4,$ $rypk_1, rypk_2, rypk_3, rypk_4,$ $rzpk_1, rzpk_2, rzpk_3, rzpk_4$	The time zone (indexes of time tick) of the top four data peaks of x, y and z axes rotation data, respectively
$p_{max}, p_{min}, p_{std}, p_{avg}$	The statistical description of the ambient air pressure data segment from the smartphone's barometer
$me_{max}, me_{min}, me_{std}, me_{avg}$	The statistical description of the magnetic field data segment from the smartphone's magnetometer
mag_{offset}	The offset of the DFT result of the magnetic field data segment
$l_{max}, l_{min}, l_{std}, l_{avg}$	The statistical description of the environment brightness from the smartphone's light sensor

where p_k is the cardinality of k-th index set, and $\|\beta^{(k)}\| = \sqrt{\sum_{j \in I_k} \beta_j}$ with I_k indicating the index set.

To apply the group lasso technique, we still use travel mode detection as an example and we categorize the used features by their source sensors obtaining five feature groups: accelerometer based features, gyroscope based features, light sensor based features, barometer based features, and magnetometer based features. Among the five groups, accelerometer and gyroscope are motion based groups. The rest feature groups are environmental based groups. In order to verify our hypothesis that the travel modes that involve more motion activities might rely on motion based features more than others modes, we first categorize the six travel modes into two classes: wheeled travel modes and unwheeled travel modes. Wheeled travel modes include biking, taking a bus, taking a subway and driving a car. Unwheeled travel modes include walking and jogging. We further split the wheeled modes into two subcategories: indoor modes (taking a bus, taking a subway, taking a bus) and outdoor mode (biking). The reason for the wheeled/unwheeled categorization is that the user's body is moving more drastically while walking and jogging. This would reflect on the smartphone sensor readings such as acceleration and rotation. As for indoor/outdoor categorization, the intuition is that the readings from environmental sensors (e.g. magnetometers, light sensor, barometers) might show different patterns due to the ambient environment difference between indoor travel tools and at outdoors. The results are shown in Figs. 3.2 and 3.3.

In Figs. 3.2 and 3.3, we input all the five feature groups into the group lasso model, and tune the λ parameter in Eq. (3.1). We can choose more groups of features as λ becomes smaller. The vertical dash lines in the figures are the moments where new group(s) of features join the whole features used for classification. Each line of red letter(s) are the total sensor group(s) in use with a blue dash arrow pointing to the dash line showing the moments when these group(s) of features start in use. In the figures, 'A', 'R', 'B', 'P', and 'M' indicate the accelerometer, gyroscope, barometer, light sensor, and magnetometer, respectively. For example, in the first

Fig. 3.2 Solution path of a linear model with group lasso applied: accuracy path for wheeled and unwheeled mode classification

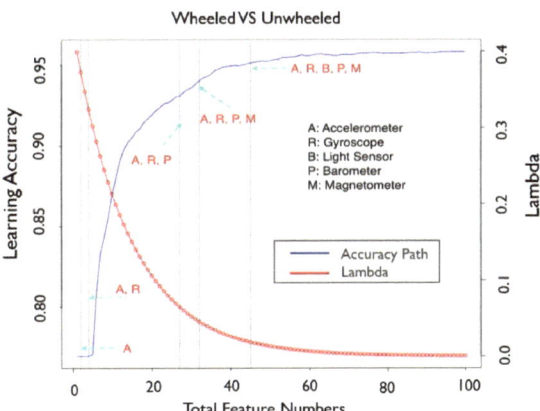

Fig. 3.3 Solution path of a linear model with group lasso applied. (**a**) Accuracy path for walking and jogging mode classification. (**b**) Accuracy path for indoor and outdoor mode classification

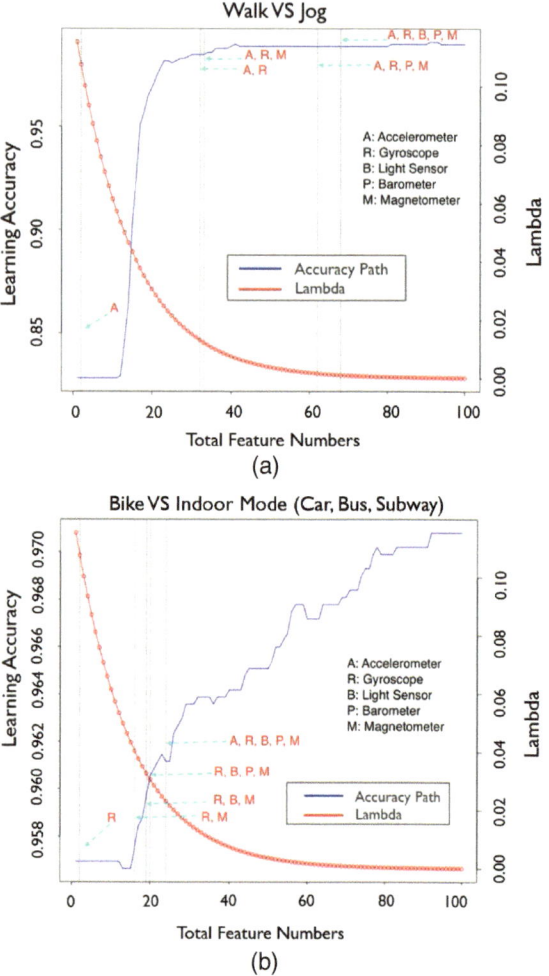

figure, the first vertical dash line is 'A' and the second is 'A, R'. This means that the gyroscope features are selected by the group lasso at the moment of the second dash line, and these features are added for classification. From the figures, we can conclude that: (1) accelerometer and gyroscope together are indicative in terms of classifying between unwheeled and wheeled travel modes; (2) unwheeled travel modes (walking and jogging) can reach a higher classification accuracy with features only drawn from accelerometer readings; (3) inside wheeled travel modes, features drawn from gyroscope and magnetometer are important to classify indoor/outdoor modes. This gives us a hint in dynamically selecting the sensors in the travel mode detection task, as well as other mobile data mining tasks.

3.4 Summary

In this chapter, we first discuss the slide-window based data segmentation method, and then introduce the feature extraction method based on the segmented data. Next, we take the travel mode detection problem as an example, illustrate what features can be extracted for this task, and show how to select the subset of features based on the sensors. The conclusion holds for most mobile data mining tasks, i.e., not all features are useful for a specific task.

Chapter 4
Hierarchical Model

Abstract After the features are ready to use, we start to present the learning models for mobile data mining applications. The learning models mainly need to consider two challenges: energy-saving and personalization. In this chapter, we present a hierarchical learning framework for mobile data mining tasks with the goal of energy-saving. Specially, we illustrate the idea with the travel mode detection task. We classify the six modes into wheeled modes and unwheeled modes, where the wheeled modes include outdoor modes (biking) and indoor modes (taking a subway, driving a car, and taking a bus), and the unwheeled modes include walking and jogging. Corresponding to the classification, the hierarchical model consists of three layers. It is based on the results of group feature analysis in the previous chapter. That is, not all sensor data are required for a certain task. For example, we find that only wheeled modes require the full sensor data while the majority of the sensors (except for accelerometer and gyroscope) are turned off in other cases.

4.1 Problem Description

The hierarchical model takes the extracted features from smartphone sensor data as the input, and learns a classification model that can predict the travel mode for the current user. One challenge here is that the continuous operation with multiple sensors could lead to the smartphone's battery discharge after a few hours. Thus, despite the general goal of high classification accuracy, learning with smartphone sensor data also demands the model to be less battery draining on the smartphone side. In this chapter, we explore the solution of a hierarchical model that provides a dynamic sensor selection mechanism to address this problem. The basic idea is to select a subset of sensors to work for the travel mode detection problem. An example is shown in Fig. 4.1 where we compare the accelerometer readings when a user is taking a bus and when s/he is walking. We can see that the acceleration is very discriminative as the readings are more fluctuated with much higher magnitude in walking. Therefore, we consider to use this accelerometer sensor only for dividing these two travel modes.

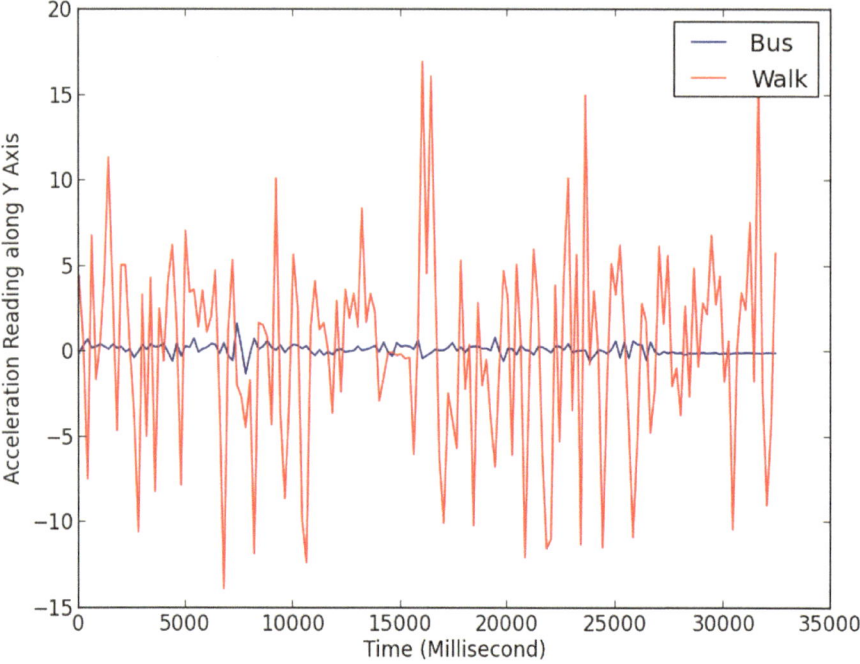

Fig. 4.1 Acceleration along the y-axis: walking v.s. taking a bus

4.2 A Hierarchical Framework

Based on the above intuition, we propose a hierarchical categorization of the six travel modes based on the results of group feature analysis in the previous chapter. The hierarchical structure separates the travel mode into layers of subcategories, as shown in Fig. 4.2. It contains three layers in total. In the first layer, the travel modes is categorized into wheeled travel mode (bus, train, bike and car) and unwheeled travel mode (walking, jogging). The intuition is that during a wheeled traveling, the body movement is less drastic than the unwheeled traveling, which can be reflected based on the accelerometer readings. In the second layer of wheeled travel mode, it splits into two sub-categories: indoor mode (bus, car, subway) and outdoor mode (bike), based on the observation that barometer readings could be different between indoor and outdoor modes. The third layer further classifies the indoor modes into bus, car, and subway. Each layer has its own classifier.

With the above hierarchical framework, along with the feature analysis that not all travel modes need all the features drawn from the readings of all sensors, we could dynamically select a subset of sensors based on current travel modes need to be classified. Take Fig. 4.3 as an example, where we have four learning models

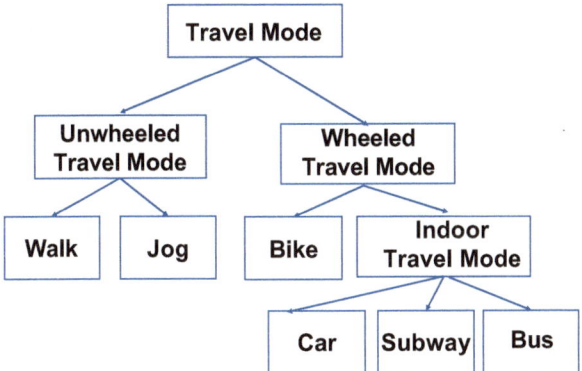

Fig. 4.2 Hierarchical categorization of travel modes

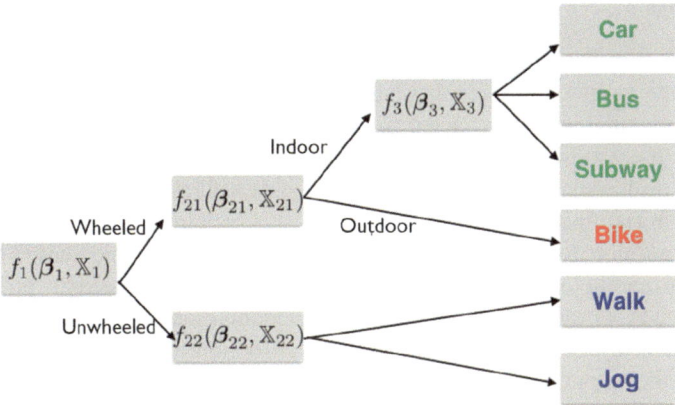

Fig. 4.3 Hierarchical learning workflow

for the hierarchical categorization. Suppose f_1 is the model at wheeled/unwheeled level; f_{21} is the model at indoor/ourdoor level within wheeled subcategory; f_{22} is the model for walking/jogging classification at unwheeled subcategory; f_3 is the model for car/bus/subway classification at the indoor subcategory of wheeled mode. The classification results will be calculated combining all four models:

$$y = \mathbb{1}_{f_1=1} \cdot \mathbb{1}_{f_{21}=1} \cdot f_3 + \mathbb{1}_{f_1=1} \cdot \mathbb{1}_{f_{21}=-1} + \mathbb{1}_{f_1=-1} \cdot \mathbb{1}_{f_{22}}$$

For f_1, we can directly use acceleration data for the classification. Even a rule-based binary classifier can be used in this level. Compared to other models, the rule-based method is less computational expensive, and more suitable for battery saving especially when only first-level classification is required. For the other

---

Algorithm 4.1 Hierarchical framework with dynamic sensor selection

---

1: Input: The learning models for each level: f_1, f_{21}, f_{22} and f_3.
2: Output: S_1, S_{21}, S_{22} and S_3, the sensor groups selected with group lasso. S_{all} is the whole sensor set of current smartphone.
3: **for** $t = 1 : T$ **do**
4:       Calculate $y_{1_t} = f_1(\mathbf{w}_1, \mathbf{x_t})$
5:       **if** $y_{1_t} == 1$ **then**
6:             Turn on sensors: $S_1 \cup S_{21}$
7:             Calculate $y_{21_t} = f_{21}(\mathbf{w}_{21}, \mathbf{x_t})$
8:             **if** $y_{21_t} == 1$ **then**
9:                   Turn on sensors: $S_1 \cup S_{21} \cup S_3$
10:             **end if**
11:       **else**
12:             Turn on sensors: $S_1 \cup S_{22}$
13:       **end if**
14: **end for**

---

models, we use the features selected by group lasso. Note that group lasso helps to select the group of features that belong to each sensor. For example, the group lasso returns the features extracted from the acceleration data for f_{22}. The reason is that the body movement is more drastic during jogging, making the acceleration data very discriminative. Note that f_{21} and f_{22} as well as f_{22} and f_3 are mutually exclusive in learning process.

The algorithm for the hierarchical framework with dynamic sensor selection is summarized in Algorithm 4.1. As we can see, we only need to turn on the sensors when we are moving to the corresponding subcategories. Additionally, the hierarchical structure provides the users an option to choose the level of the details of the output. For example, some transportation surveys are only interested in distinguishing wheeled travel modes from unwheeled travel modes. In this case, only the first-level classification model would be applied.

4.3 Experimental Evaluations

The experiments are designed to compare the performance between the hierarchical model and the general model. For all the models except f_1, we use a multi-class SVM classifier with one-against-all method. The hierarchical model works following the procedure shown in Algorithm 4.1.

Before comparing the hierarchical model with the general model, we first conduct some experiments to estimate the sensors' battery consumption on several phone models (iPhone 5, 6, and Samsung Galaxy S4). The power usage is estimated at different sampling frequencies when the target sensor is turned on and the other sensors and related Apps are all turned off. We combine the results of our experiments with conclusions from other literature [24, 30, 55] to give the estimated sensors battery consumption of the five major sensors. Instead of giving the exact

Table 4.1 Sensors' power consumption

Sensor in use	Power consumption ratio
Phone idle	1
Light sensor (1 Hz)	1
Barometer sensor (1 Hz)	1.02
Accelerometer (5 Hz)	1.59
Magnetometer (5 Hz)	1.45
Gyroscope (5 Hz)	1.82

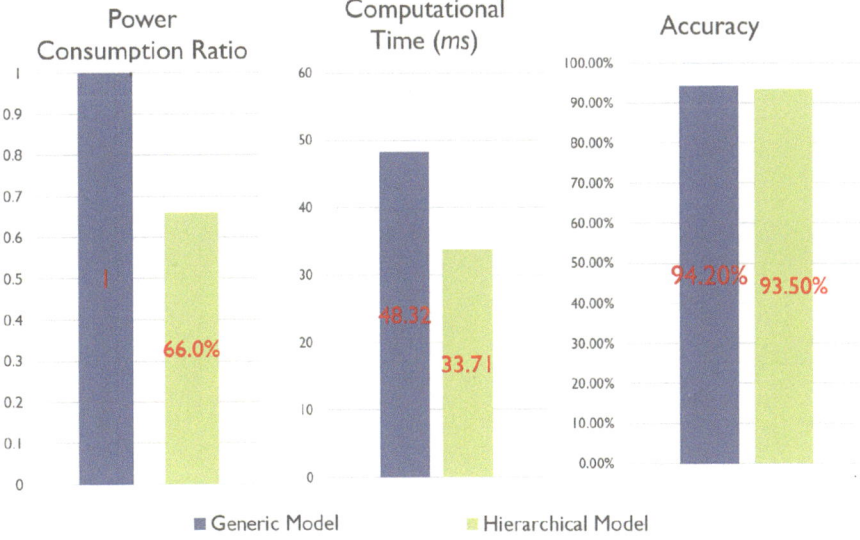

Fig. 4.4 Performance comparison of hierarchical model and general model

numbers, which might be different from phone to phone, we use the ratio of battery consumption by comparing to the battery consumption when the phone idles and the result is shown in Table 4.1.

Next, we evaluate the performance of the hierarchical model by comparing the average sensors in use, the total calculation time, and the prediction accuracy. The results are shown in Fig. 4.4. We can see that the hierarchical model is comparable to the general model in prediction accuracy. Meanwhile, it outperforms the general model in total sensors in use as well as the time spent for computing. We also estimate the power consumption based on the conclusion in Table 4.1. The result shows that the hierarchical model works well in improving the battery efficiency on smartphones, and it uses only 66% of the battery consumption compared with the general model.

4.4 Summary

In this chapter, we have introduced a hierarchical framework for mobile data mining tasks. The basic idea stems from the feature analysis results that only part of the sensors can achieve sufficiently accurate predictions, especially for certain classification tasks. Based on this idea, hierarchical models can save the energy of mobile apps by using a subset of sensors during each classification in the hierarchical layers.

Chapter 5
Personalized Model

Abstract Another challenge for mobile data mining tasks is personalization. On one hand, we usually do not have plenty labels for a certain user, and thus we need to borrow the labeled data from other users. One the other hand, different users tend to behavior differently and thus have different patterns of sensor readings, making data borrowing non-trivial. In this chapter, we introduce a personalized treatment for mobile data mining tasks. The basic idea can be divided into two stages. In the first stage, for a given target user, we select and borrow some data from the users whose labeled data are already collected in the database. In the second stage, we reweight the borrowed data and use them as the training data for the target user. The proposed method is able to estimate the sample distributions, and then reweight the samples based on the estimated distributions so as to minimize the model loss with respect to the target user's data.

5.1 Problem Description

Existing mobile data mining methods tend to train a generic classifier for all users, while the behaviors may differ greatly from one user to another. This limitation indicates that a personalized treatment may result in better prediction accuracy. However, following personalized treatment, another limitation stems from the *label sparsity*. That is, there may be a very limited set of labels for a certain user, making the trained classifier less effective. To handle these challenges, we need to borrow the data from other users whose data are pre-collected and labeled. The two key problems here is to determine whose data to borrow, and how to use the borrowed data.

Before presenting the proposed personalized model, we first introduce some terminology and notations used in this chapter.

Base Data We use base data B to indicate the pre-collected smartphone data. Each user in the base data has a collection of data samples, and each data sample is represented as $\{\mathbf{x}, y\}$, where $\mathbf{x} \in \mathbb{R}^p$ is a p-dimensional feature mapped from the raw data and $y \in \{y_1, y_2, ..., y_m\}$ is one of the m travel mode labels.

Y. Yao et al., *Mobile Data Mining*, SpringerBriefs in Computer Science,
https://doi.org/10.1007/978-3-030-02101-6_5

31

Target Data The target data T is the data from the user that we aim to predict the travel modes for. We assume that there is a small portion of available labels in T. We further denote the data distribution of the feature space in T as P_T.

Source Data The source data S is a subset of the base data B. It is selected from B with the aim to learn a personalized model for the user in target data T. The data distribution of the feature space in S is denoted with P_S.

Based on the above terminology, the personalized travel mode detection problem can be described as follows. We are given (1) a collection of existing users' travel data B where all the data samples are labeled, and (2) the target user's travel data T where only a small portion of data samples are labeled. The goal is to find (1) a subset of source data S from B with respect to T; (2) the weight for each data sample in S, and (3) the labels for the unlabeled data in T.

5.2 The Personalized Approach: Overview

Next, we take the travel mode detection problem as an example and illustrate the personalized model. We first present the overview of the proposed personalized model. As mentioned above, due to the diversity in individuals' travel behaviors, simply learning a generic model from the labeled data may not result in good performance. Additionally, learning a personalized model for a target user may face the issue of insufficient labeled data which would cause a high variance of the trained travel mode detector. To this end, our idea is to transfer the knowledge from the pre-collected data, and adapt the knowledge to the target user's travel behaviors. The overview of the proposed method is shown in Fig. 5.1.

As we can see from Fig. 5.1, given the input, there are two major steps:

1. *Sample selection*: what kind of data to *borrow* for a given target user in T?
2. *Sample reweighting*: how to *reweight* the borrowed data for the target user?

For the first step, the basic idea is to select samples from B that are most similar to the small portion of labeled data in T. Here, one of the key challenges is to quantify the similarities between users in terms of their travel behaviors. The similarity reflected in the data would be similar data range and fluctuation in motion related sensor readings, or similar environment parameters like ambient air pressure, etc. Meanwhile, we observed that the non-sensor factors such as gender, city of travel events, and smartphone types also play important roles in the user similarity computation. For example, the phone model as well as the operating system defines how sensitive the sensors are, which directly affects the data readings. We will explore both sensor factors and non-sensor factors in our user similarity computation. Moreover, since it is possible two users having similar jogging patterns might act quite differently in driving, the similarity computation is conducted within each travel mode.

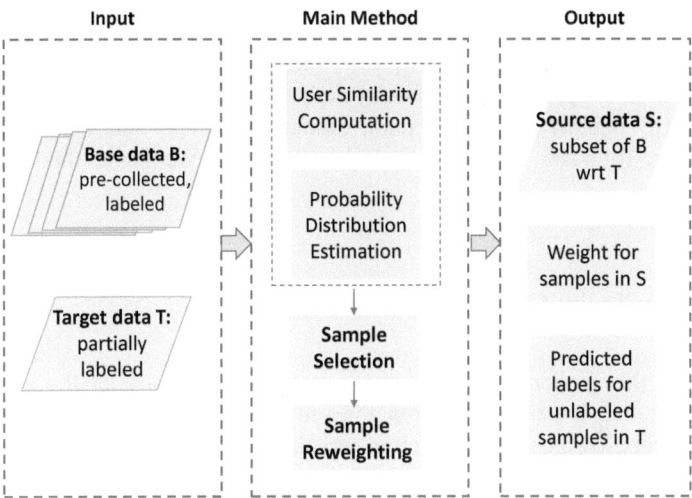

Fig. 5.1 Overview of the personalized model

For the second step, the key challenge is to reweight the selected samples (i.e., source data S) from the first step, so as to minimize the mean distance between the data in S and the labeled data in T. In our model, we adopt Kernel Mean Matching (KMM) method [27, 56] for this step. KMM is used under the situations when the distributions of training data and test data do not match, and it proposes to minimize the loss of learning by reweighting the samples in the training set. Formally, KMM aims to find the suitable weights $\beta(\mathbf{x})$ for the minimization of the following formulation

$$R[P_T, \theta, l(\mathbf{x}, y, \theta)] = \mathbf{E}_{(\mathbf{x},y)\sim P_T}[l(\mathbf{x}, y, \theta)]$$

$$= \mathbf{E}_{(\mathbf{x},y)\sim P_S}[\beta(\mathbf{x})l(\mathbf{x}, y, \theta)] \qquad (5.1)$$

where P_T and P_S are the estimated distributions. The key finding of KMM is that when $\beta(\mathbf{x}) = \frac{P_T(\mathbf{x},y)}{P_S(\mathbf{x},y)}$, the kernel mean difference between the test data and the reweighted training data is the minimum. In our problem setting, we also match the distributions within each travel mode.

The output of the proposed method is the selected and reweighted data S from B with respect to T, as well as the predictions which can be made based on S and T. In the following section, we will present the details of the two major steps.

5.3   The Personalized Approach: Details

In this section, we present the details of the proposed personalized model, including similarity computation, distribution estimation, sample selection, and sample reweighting, followed by some algorithm analysis. We name this method as *PerTMoD*.

5.3.1   Similarity Computation

To analyze the similarity of two users' data, we first consider the sensor factors/features that are extracted from the sensor data. The used sensors include accelerometers, barometers, magnetometers, and gyroscope (see the experimental section for more details). Based on the data readings, we define the following two metrics.

- **Data Center**. A data center is associated to a specific user and a specific travel mode. We denote a data center associated with user i and travel mode k by $C_i^{(k)}$ ($k = 1, 2, ..., m$)

$$C_i^{(k)} = < \bar{x}_{i,1}^{(k)}, \bar{x}_{i,2}^{(k)}, ..., \bar{x}_{i,p}^{(k)} > \tag{5.2}$$

  where $\bar{x}_{i,j}^{(k)}$ is the mean of the j-th features that is associated to user i and mode k.

  For user i, the data center profile is a vector of length m: $< C_i^{(1)}, C_i^{(2)}, ..., C_i^{(m)} >$. If this user does not have certain mode k in his/her travel data, we will skip this value in calculation.

- **Distance**. The travel data distance is defined as the Euclidean distance of two data centers of the same mode. We denote it by $D_{i,i'}^{(k)}$.

$$D_{i,i'}^{(k)} = \| C_i^{(k)} - C_{i'}^{(k)} \|_2 \tag{5.3}$$

For non-sensor factors (which is related to a user's personal information), we consider the following four factors: gender, phone model, city of travel, and the season when travel happened. We use A to denote the non-sensor similarity matrix and each entry $A_{i,i'}$ is defined as

$$A_{i,i'} = \sum_l a_{i,i'}^{(l)} \tag{5.4}$$

with

$$a_{i,i'}^{(l)} = \begin{cases} 1 - \xi, & \text{if user } i \text{ and } i' \text{ have the} \\ & \text{same value in } l\text{-th factor,} \\ \xi, & \text{otherwise.} \end{cases}$$

where $0.5 \le \xi < 1$ is a parameter to control the importance of non-sensor factors.

Based on the Eqs. (5.3) and (5.4), we define the similarity score between two users under the same label k as

$$o_{i,i'}^{(k)} = \frac{A_{i,i'}}{D_{i,i'}^{(k)}} \tag{5.5}$$

As we can see from the above equation, the higher the non-sensor similarity and the smaller the sensor distance, the more similar the two users are. For parameter ξ, smaller value of ξ indicates higher impact of the non-sensor factors.

For a given label k, we can generate its similarity matrix $O^{(k)}$ with its entries as defined in Eq. (5.5). The data samples from the top-n similar users form the candidate set S_C for the source data S.

5.3.2   Distribution Estimation

Another essential component of our method is the distribution estimation. For any data sample $\mathbf{x}_i$ in the feature space, we define its potential to another data point $\mathbf{x}_j$ in the same space as $V(\mathbf{x}_i, \mathbf{x}_j)$,

$$V(\mathbf{x}_i, \mathbf{x}_j) = \frac{1}{\|\mathbf{x}_i - \mathbf{x}_j\|_2^2 + \gamma} \tag{5.6}$$

where γ is a variable that defines the sharpness of the potential distribution. The more close two data points are, the higher the potential is between them. Based on the potential, we can define the probability distribution of any data S under a certain label by the density function:

$$P_S(\mathbf{x}|y) = N_S \sum_{\mathbf{x}_j \in S} V(\mathbf{x}, \mathbf{x}_j) \tag{5.7}$$

$$= N_S \sum_{\mathbf{x}_j \in S} \frac{1}{\|\mathbf{x} - \mathbf{x}_j\|_2^2 + \gamma}.$$

where N_S is the normalizer for data S so that $\sum_{\mathbf{x}_i \in S} P_S(\mathbf{x}_i) = 1$. With this estimation, we can use the mean matching process to assign weights such that the distributions of two sets of data samples are close to each other.

5.3.3 Sample Selection

Simply using the data samples from most similar users (i.e., the output of Sect. 4.1) may cause the label imbalance problem. Here, we tackle this problem by defining a filter function.

The data from top-n similar users will form a ball such that all the data samples fall inside the sphere of the ball. We define a ball G whose center is $\mathbf{c}$ and whose radius is r. Using all the data from top-n similar users means that r should satisfy the following inequality

$$r \geq \sup_{\mathbf{x} \in T}\{\|\mathbf{x} - \mathbf{c}\|_2\}.$$

In order to balance the data between S and T, we choose a smaller r to serve as a filter. That is, we define the following filter function $F_G(\mathbf{x}) : S \cup T \mapsto \mathbb{R}$

$$F_G(\mathbf{x}) = \begin{cases} 1, & \mathbf{x} \in G \\ 0, & \text{otherwise.} \end{cases} \tag{5.8}$$

The filter depends on the choice of r and it filters out the points outside the ball G in the feature space. We apply the filter on the candidate set S_C to obtain S. By doing this, we make sure that the points in T are contained in the range of S. The updated P_S is given by

$$P_S'(\mathbf{x}|y) = \frac{P_S(\mathbf{x}|y) \cdot F_G(\mathbf{x})}{\sum_{\mathbf{x}' \in S} P_S(\mathbf{x}'|y) \cdot F_G(\mathbf{x}')} \tag{5.9}$$

We do not need to apply F_G on P_T because T is already inside the scope of G by definition.

5.3.4 Sample Reweighting

So far, we have the source data S selected. We next move to the sample reweighting via distribution matching. A prerequisite of distribution matching is that the two distributions are overlapped in their domain of definition, which is guaranteed by our sample selection method. Basically, the weight is assigned towards each $\mathbf{x}$ in S according to its relationship to the samples in T. Here, we directly show the computation of weights in the following equation and leave the proof for the next subsection.

Algorithm 5.1 The *PerTMoD* algorithm

1: Input: base data B and target data T
2: Output: source data S and its weight vector β_S
3: Parameters: top-n most similar users in sample selection, the similarity score parameter ξ in Eq. (5.5), and the potential parameter γ in Eq. (5.6)

4: $S \leftarrow \{\}$;
5: $\beta_S \leftarrow \{\}$;
6: **for all** $k \in 1 : K$ **do**
7:     Calculate the similarity matrix $O^{(k)}$ between B and labeled data in T by Eq. (5.5);
8:     Select n most similar users in current travel mode k to form S_C;
9:     Apply the filter in Eq. (5.8) on S_C to obtain S_k;
10:     Calculate the weight $\beta_{Sk}(\mathbf{x})$ for $\mathbf{x} \in S_k$ by Eq. (5.10);
11:     $S \leftarrow S \cup S_k$;
12:     $\beta_S \leftarrow \beta_S \cup \beta_{Sk}$;
13: **end for**
14: **return** S and β_S;

$$\beta_y(\mathbf{x}) = \frac{P_T(\mathbf{x}|y)}{P_S'(\mathbf{x}|y)} \tag{5.10}$$

$$= \frac{\sum_{\mathbf{x}_j \in T} \frac{N_T}{\|\mathbf{x}-\mathbf{x}_j\|_2^2+\gamma}}{\sum_{\mathbf{x}_j \in S} \frac{N_S}{\|\mathbf{x}-\mathbf{x}_j\|_2^2+\gamma}}$$

If T and S are the same set, we have $\beta_y(\mathbf{x}) = 1$ for any $\mathbf{x}$.

The overall algorithm for our *PerTMoD* is summarized in Algorithm 5.1. In the algorithm, we compute the similarities between data samples in B and T, select the samples from B, and then compute the weights of the selected samples, respectively. The above steps are conducted in each travel mode. We apply the output of the algorithm on the weighted SVM as shown in Eq. (5.11) to predict the labels for unlabeled data samples in T.

$$\begin{aligned} \underset{\mathbf{w},\xi}{\text{minimize}} \quad & \frac{1}{2}\|\mathbf{w}\|^2 + C \sum_{i=1}^{n_{tr}} \beta(\mathbf{x}_i)\xi_i \\ \text{subject to} \quad & y_i \langle \mathbf{w}, \mathbf{x}_i \rangle \geq 1 - \xi_i \\ & \xi_i \geq 0, i = 1, 2, \ldots \end{aligned} \tag{5.11}$$

5.3.5   Algorithm Analysis

Here, we briefly analyze the proposed personalized algorithm (Algorithm 4.1).

The proposed algorithm is built upon the Kernel Mean Matching [27]. It is proved that the loss function in Eq. (5.1) is bounded given that the solution $\beta(\mathbf{x})$ minimizes the following problem

$$\min_{\beta} \quad \|\mu(P_T) - \mathbf{E}_{x \sim P_S(\mathbf{x})}[\beta(\mathbf{x}) \cdot \mathbf{x}]\|$$

$$\text{subject to} \quad \beta(\mathbf{x}) > 0, \tag{5.12}$$

$$\mathbf{E}_{x \sim P_S(\mathbf{x})}[\beta(\mathbf{x})] = 1.$$

In the following lemmas, we show that the proposed sample weight estimation in Eq. (5.10) also meets the condition in Eq. (5.12).

Lemma 5.1 *In each labeled class y_i, given the construction of $P_T(\mathbf{x}|y_i)$ and $P_S(\mathbf{x}|y_i)$, $\beta_{y_i}(\mathbf{x}) = \frac{P_T(\mathbf{x}|y_i)}{P_S(\mathbf{x}|y_i)}$ guarantees the minimum of the following problem:*

$$\min_{\beta_{y_i}} \quad \|\mu(P_T|y_i) - \mathbf{E}_{\mathbf{x} \sim P_S(\mathbf{x}|y_i)}[\beta_{y_i}(\mathbf{x}) \cdot \mathbf{x}]\|$$

$$\text{subject to} \quad \beta_{y_i}(\mathbf{x}) > 0, \tag{5.13}$$

$$\mathbf{E}_{x \sim P_S(\mathbf{x}|y_i)}[\beta(\mathbf{x}|y_i)] = 1.$$

Proof First, since we focus on the importance of samples within each label class y_i with respect to the data samples of the same class in T, for a fixed y_i, we have

$$\beta_{y_i}(\mathbf{x}) = \frac{P_T(\mathbf{x})}{P_S(\mathbf{x})} = \frac{P_T(\mathbf{x}|y_i)}{P_S(\mathbf{x}|y_i)} \tag{5.14}$$

In Sect. 5.3.3 the filter function applied on P_S in Eq. (5.8) guarantees that range of data S in the feature space contains T, thus $P_T \ll P_S$. This guarantees the validity of Eq. (5.13). Then, by the definitions of P_S and P_T, the calculation of $\beta_{y_i}(\mathbf{x})$ in Eq. (5.14) satisfies the constraints and gives Eq. (5.13) the value of 0, which completes the proof.   □

Based on Lemma 5.1, we have that the minimization within each class label guarantees the overall minimization of the distribution matching, as shown in the following lemma.

Lemma 5.2 $\beta_y(\mathbf{x})$ *as defined in Eq. (5.10) guarantees the minimization of Eq. (5.12).*

5.4 Experimental Evaluations

In this section, we present the experimental setup, and discuss the experimental results.

5.4.1 Experiment Setup

In the experiments, we select 10 users as the target users and use their data as target data T. For each target user, we hide $(1 - \alpha)$ of its labels. We assume that there are a small portion of available labels, which means α is usually very small (e.g., 1%). We use the data from other users as base data B. The proposed method is applied on the target data and base data. Then, we report the prediction accuracy of each user when this user is selected as the target user.

To evaluate the effectiveness of *PerTMoD*, we compare the following methods:

- *Generic*. The generic method ignores user personalization and trains the classifier on all available labeled data.
- *Basic*. The basic method indicates the case when only the labeled data in T is used as training data.
- *Random*. Personalized prediction with transferred data randomly selected.
- *Selection*. Personalized prediction where we use the transferred data samples in S without reweighting.
- *R-Selection*. Personalized prediction where we add the data samples that are not in S as training data.
- *R-Reweighting*. Personalized prediction when we add sample reweighting on the *R-Selection* method.
- *PerTMoD*. The proposed personalized travel mode detection method.

5.4.2 Experiment Results

First, we compare the effectiveness of different methods, and the result are shown in Fig. 5.2 where the x-axis is the user id, and the y-axis is the prediction accuracy. The label portion is set as 2% (i.e., $\alpha = 2\%$) in the figure. The three parameters of the *PerTMoD* method are tuned via cross validation. In particular, the reported results are under the parameter setting $n = 3, \xi = 0.2, \gamma = 0.003$.

We can first observe from the figure that, the *PerTMoD* method generally outperforms all the compared methods. For example, averaging over the 10 users, *PerTMoD* improves the *Generic* method by 22.3% in terms of absolute prediction accuracy. This result validates the effectiveness of the personalization treatment in *PerTMoD* for the travel mode detection problem. Additionally, *PerTMoD* is averagely 5.2% better than the *Basic* method. This result indicates the usefulness of transferring knowledge from existing data. Second, we can observe that *Selection* is better than the *Random* method. This result verifies the effectiveness of the proposed data selection method (i.e., the first stage) in *PerTMoD*. Third, *PerTMoD* is better than the *Selection* method. Moreover, we find that applying the reweighting method on the data that is not in S can still improve the prediction accuracy (i.e., *R-Reweighting* is better than *R-Selection*). This result further verifies the usefulness of sample reweighting (i.e, the second stage in *PerTMoD*).

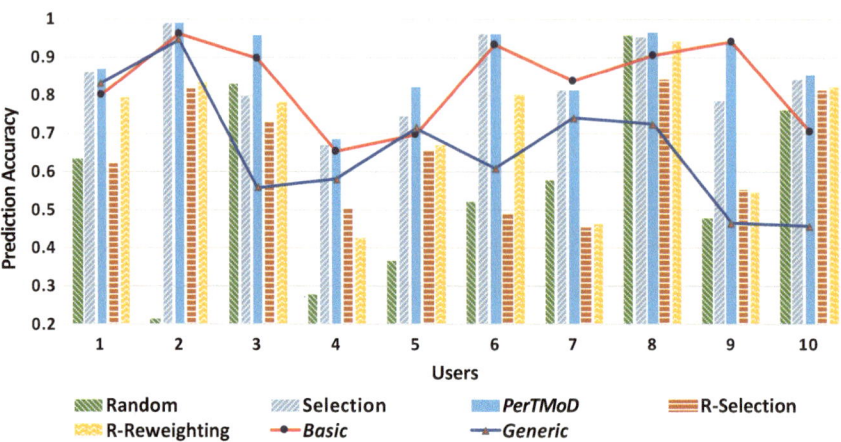

Fig. 5.2 Performance comparison of personalized model and other models

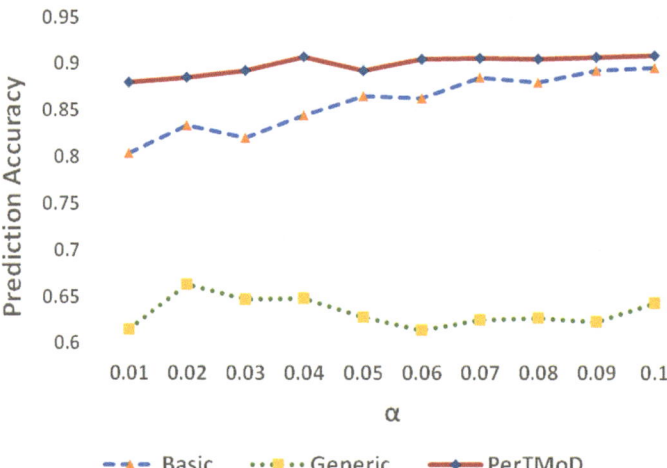

Fig. 5.3 Performance comparison of personalized model when α varies

Next, the effectiveness results when the portion (α) of available labels in T varies are shown in Fig. 5.3. In the figure, we vary α from 1% to 10%, and report the average prediction accuracy over all users. As we can see, *PerTMoD* consistently outperforms the *Generic* method and the *Basic* method, indicating the usefulness of both the proposed personalization treatment and the proposed knowledge transfer method.

5.5 Summary

In this chapter, we tackle the personalization challenge of mobile data mining tasks. To handle the this challenge as well as the related label sparsity challenge, we propose a two-step personalized treatment. The first step determines the users that we can borrow data for the current user, and the second step adapts these borrowed data for the prediction of the current user.

Chapter 6
Online Model

Abstract Model updating is important for mobile data mining tasks. When more labeled data are available for a given target user, the mobile data mining algorithm should incorporate such data in order to provide more accurate services for the user. However, directly re-training the model with both the old and the new data would be resource consuming especially for mobile applications. A more desired way is to incrementally update the model in an online fashion. In this chapter, we introduce an online model for the mobile data mining tasks. The online model is orthogonal with the hierarchical model and personalized model. The basic idea is to adopt the stochastic sub-gradient descent method and updates the learning models with a small portion of new data.

6.1 Problem Description

In this chapter, we focus on the problem of temporal model adaption. That is, given a user-specific model, how do we update it with new labeled data? Still take the travel mode detection problem as an example. The common scenario for this problem is when a user provides the feedback (i.e., the correct label) to his/her current travel mode. With the new labeled data, we would like to update the pre-trained model. Re-training the model with the whole updated dataset will be both time and resource consuming, providing the scale of travel data volume. Additionally, the shorter the model updating process is, the better. To meet these criteria, the classification model should satisfy the following two principles:

1. The classification model can process a small batch of samples despite the data used previously. In other words, it should use online updating strategy.
2. The model updating algorithm should not put restrictions on the data scale since each update should only depend on data sensed from the current traveler's behavior.

In the following part, we introduce an online learning model that meets the above principles and it detects the travel mode with promising accuracy at the same time.

Y. Yao et al., *Mobile Data Mining*, SpringerBriefs in Computer Science,
https://doi.org/10.1007/978-3-030-02101-6_6

6.2 Online Learning

The proposed online learning method can be applied to many classification models. Here, we take support vector machine (SVM) as an example. SVM is a supervised classification tool that provides the largest margin between two hyperplanes of the classes in the multi-dimensional feature space. In a binary classification problem, we have a training set $S = \{(\mathbf{x_i}, y_i)\}_{i=1}^{m}$, where $\mathbf{x_i} \in \mathbf{R}^n$ and $y_i \in \{+1, -1\}$. The pair $(\mathbf{x_i}, y_i)$ is composed of an arbitrary input $\mathbf{x}$ and the prediction label y. To train a SVM is to find the minimizer of the following problem:

$$\min_{\mathbf{w}} \left(\frac{\lambda}{2} \|\mathbf{w}\|^2 + \frac{1}{m} \sum_{(\mathbf{x},y) \in S} \ell(\mathbf{w}; (\mathbf{x}, y)) \right) \tag{6.1}$$

where,

$$\ell(\mathbf{w}; (\mathbf{x}, y)) = \max\{0, 1 - y\langle \mathbf{w}, \mathbf{x} \rangle\} \tag{6.2}$$

We denote the objective function in Eq. (6.1) by $f(\mathbf{w})$. Subgradient descent has often been proposed to find a solution for the approximate objective functions like $f(\mathbf{w})$ [52]. A simplification for subgradient descent is stochastic gradient descent (SGD). SGD allows the update on batch gradient descent with randomly picked samples at each iteration [53]. At each iteration, the update is given by:

$$\mathbf{w}_{t+1} \leftarrow \mathbf{w_t} + \eta \nabla_{\mathbf{w}} f \tag{6.3}$$

Since the stochastic algorithm does not need to remember which examples were visited during the previous iterations, it can process examples on the fly in a deployed system [14].

In the travel mode identification problem, on the other hand, different users could behave very differently in traveling. Therefore, it is possible that certain user's behavior shows a huge difference from the data we used to train the general model. In the meanwhile, different users may need different size of new data for the model updating in order to reach a stable performance. Thus the model updating requires user-specific samples and the outputs should be user-specific parameters. In our design, we require a model updating algorithm that does not restrict the data scale and can process with a small batch of samples despite the previous data. The SGD algorithm fits our requirements well.

In particular, we adopt the Primal Estimated sub-GrAdient SOlver (Pegasos) proposed in [58] to update SVM. Pegasos works solely on the primal objective function at each iteration, and thus its running time does not depend on the entire training set size [58]. The sub-gradient of the approximation in Eq. (6.1) is then given by:

$$\nabla_t = \lambda \mathbf{w}_t - \mathbb{1}[y_{i_t} \langle \mathbf{w}_t, \mathbf{x}_{i_t} \rangle < 1] y_{i_t} \mathbf{x}_{i_t} \tag{6.4}$$

Algorithm 6.1 Online learning with Pegasos

1: Input: $\mathbf{w_0}$, the weight vector of the general model; λ, the regularization parameter.
2: Output: The updated weight vector of the learning model: $\mathbf{w_{T+1}}$
3: Parameters: T, the maximum training epoch, $\mathbf{S}$, the training set.
4: $t \leftarrow 0$
5: **while** $t < T$ **do**
6:     $\eta_t = \frac{1}{\lambda t}$
7:     choose $i_t \in \{1, 2, ..., |\mathbf{S}|\}$ uniformly at random
8:     **if** $y_t \langle \mathbf{w_t}, \mathbf{x_t} \rangle < 1$ **then**
9:         Set $\mathbf{w_{t+1}} \leftarrow (1 - \eta_t \lambda)\mathbf{w_t} + \eta_t y_t \mathbf{x_t}$
10:    **else**
11:        Set $\mathbf{w_{t+1}} \leftarrow (1 - \eta_t \lambda)\mathbf{w_t}$
12:    **end if**
13:    $\mathbf{w_{t+1}} \leftarrow \min\{1, \frac{1/\sqrt{\lambda}}{\|\mathbf{w_{t+1}}\|}\}\mathbf{w_{t+1}}$
14:    $t \leftarrow t + 1$
15: **end while**
16: **return** $\mathbf{w_{T+1}}$

---

where $\mathbb{1}[y_{i_t} \langle \mathbf{w_t}, \mathbf{x_{i_t}} \rangle < 1]$ is the indicator function that takes a value of one if $\mathbf{w}$ yields non-zero loss on the example $(\mathbf{x}, y)$. Substituting ∇_t in Eq. (6.3) with Eq. (6.4), and using learning rate $\eta_t = 1/\lambda t$, the update of $\mathbf{w}$ becomes

$$\mathbf{w_{t+1}} \leftarrow \mathbf{w_t} + \eta_t \nabla_{\mathbf{w}} \mathbb{1}[y_{i_t} \langle \mathbf{w_t}, \mathbf{x_{i_t}} \rangle < 1] y_{i_t} \mathbf{x_{i_t}} \quad (6.5)$$

Algorithm 6.1 shows the pseudocode of the online model updating algorithm. In the algorithm, we take the initially trained (user-specific) model as the input, and employ the Pegasos method to update the model with a single training instance. The input $\mathbf{w_0}$ in the model updating process is from the pre-trained model.

6.3 Experimental Evaluations

In this section, we explain the development of experiment methodology, and then analyze the results and discuss the solutions as well as existing problems. We take the best performance of the general learning models in [63] as the baseline for the performance analysis. Our goal is to verify that the online updating solution that follows the principles of short system response time and less calculation, while it performs as well as the baseline. To begin with, we first raise the following questions that we aim to find the answer to each of them in following part:

- The baseline in [63] is using Bayes Net learning model. Here we demonstrate the online learning model with Pegasos (SVM). Before developing the online model, we need to fill the comparison gap between the method in [63] and the online learning method: the performance of offline learning with Pegasos (SVM) as the learning algorithm. Would the offline with Pegasos perform as well as the results in [63]? Our expectation is that the Pegasos updating with the whole dataset, as offline learning, should achieve similar performance in classification accuracy as in [63].

- We would further promote the learning algorithm to online mode that updates the pre-trained classification model with a single instance. Based on the expected results of the first question, would the online updating give promising results compared to the offline method?
- Since the online model updates itself using a single instance, would it excel the offline training mode in computation time?

In the following, we first compare the baseline with both online and offline updating models. Then, we present the performance results in terms of (1) energy consumption, (2) response time, and (3) the amount of data needed for training an adaptive model for smartphones users.

6.3.1 Online Learning vs. Offline Learning

To compare the online updating strategy with the traditional offline training method, we use Pegasos batch updating method in the offline training and take the whole training set as the batch size. In the online mode, we update the model with each single instance. In the offline mode, the dataset used to train a new model is the combination of the existing dataset and the new instance (the old model is abandoned). Figure 6.1 shows the process of online updating and offline updating. In order to compare the performance of online learning with offline learning, we repeat the offline learning with the whole updated dataset every time after the new data sample comes in. By doing so, we guarantee that the offline model we compare with is trained with the same data as our online model.

We compared the two updating process in prediction performance and time cost. In both experiments, we use the data collected with Android phones in winter.

The baseline we compare to is the best result in [63] using Bayes Net Model. Figure 6.2 shows that both online updating and offline updating using SVM with subgradient descent solver achieve promising results. The accuracy of the offline learning model is as good as the baseline, this answers our first question. The online updating model begins with lower accuracy: 65% recall and 75% precision in prediction. The performance improves significantly after about 50 iterations. And the accuracy of online and offline updating model converge as more data are used for training. This answers the second question. Figure 6.3 shows the time cost of online and offline updating. The time cost of online updating is relatively stable and accumulated to less than 0.01 s for nearly 250 iterations (data generated in about 53 min) of updating, while the time cost of offline updatingincreases faster and it

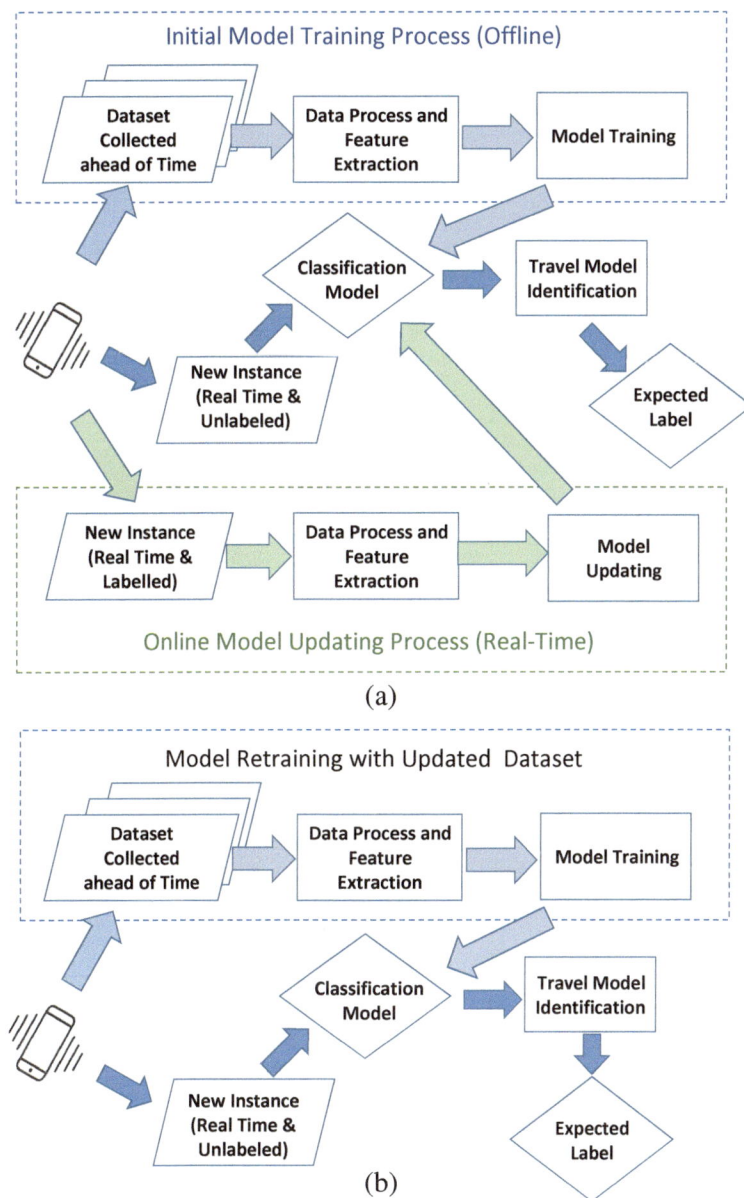

Fig. 6.1 Online v.s. offline learning process. (**a**) Online learning, (**b**) Offline learning

reaches 85 s around the 250th iteration. Together with the effectiveness results, we can conclude that the online updating is superior to the offline model. So far we have answered all three questions from the beginning of this section, and the principle 1 of online updating strategy is guaranteed.

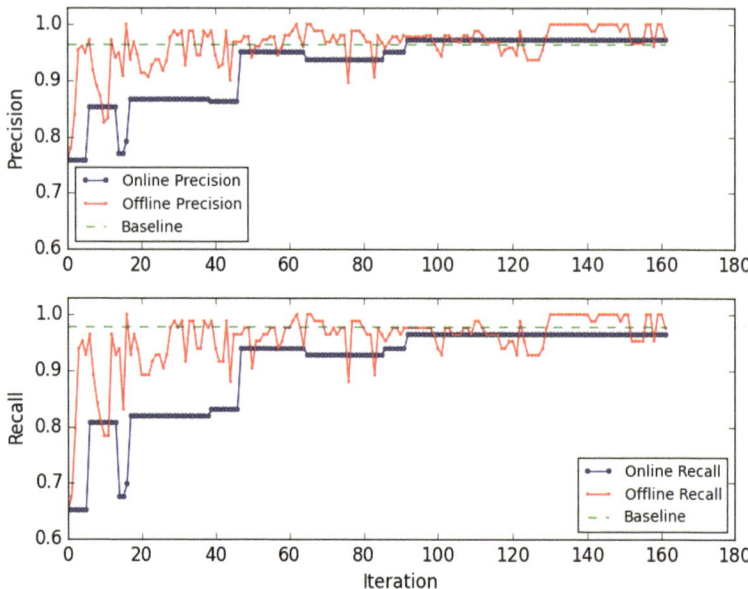

Fig. 6.2 Performance comparison of online model and offline model

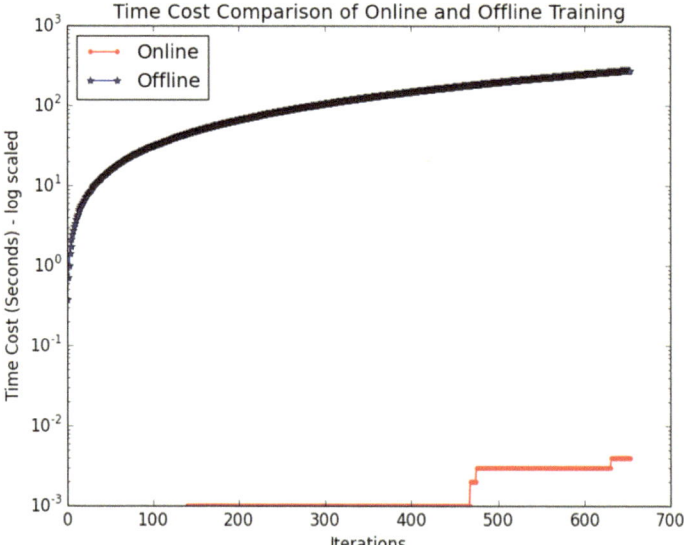

Fig. 6.3 Time cost of online model and offline model

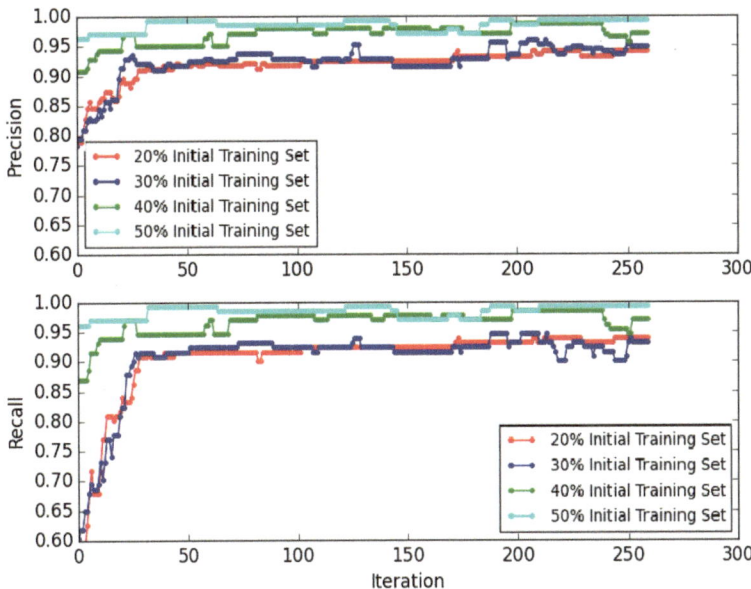

Fig. 6.4 Performance with initial training sets of different size

6.3.2 How Much Data Are Enough to Train an Initial Model?

In machine learning problems in general, the size of training dataset is also critical. Smaller dataset usually results in less accurate models. In this paper, since the model is updated with new data samples, the whole system performance should less depend on the initial training set. In order to see how much data are necessary to train the initial model, we did an experiment with different initial training sets (we take 20%, 30%, 40%, 50% of the whole training set as the initial training set and the rest as the add-up training set). Figure 6.4 shows the performance comparison with different initial training sets. It is found that smaller initial training datasets have lower accuracies at the beginning. However, as the model is updated with new coming instances, the accuracies increase and tend to converge. The experiment result reaches as high as 90% predicting precision even if only 20% of the data are used as initial training set. This result means the system meets the second principle so far as it promises a quick start for the online learning model.

6.3.3 Combining with the Hierarchical Model

Finally, we demonstrate the system performance by combining the online model with the hierarchical model. The confusion matrix of the experiment result is in

Table 6.1 Classification results

First layer	Accuracy	Second layer	Confusion matrix			
Unwheeled modes	96%	Classified as	Walk	Jog		
		Walk	2523	39		
		Jog	54	260		
Wheeled modes		Classified as	Bike	Bus	Subway	Car
		Bike	1241	0	0	1
		Bus	2	159	23	621
		Subway	49	42	4862	115
		Car	170	54	18	16,877

Table 6.1. The first layer classification accuracy reaches 96%. The second layer classification accuracy is 96.7% for unwheeled travel modes and 95.4% for wheeled travel modes.

6.4   Summary

In this chapter, we propose an online model for detecting a user's travel mode. The basic idea is to apply stochastic gradient methods to update the model when a small amount of new data arrived. We have shown that the online model tends to have close accuracy compared the offline method while it is much faster.

Chapter 7
Conclusions

Abstract In this chapter, we first present a brief summary of the content in this book, and then discuss the interesting future directions.

7.1 Summary

Smartphone has been changing the landscape of people's daily life, and has opened the door for many interesting data mining applications. In this book, we first introduce the some typical sensors in smartphones and describe the data that can be collected by these sensors, and then discuss the key applications that mobile data have been used. After that, we present the key steps and the key challenges we face towards developing a mobile data mining task. Particularly, we take the travel mode detection problem as an example to illustrate the steps and the challenges of energy-saving and personalization. More specifically, we first develop an app that allows the collection of a multitude of sensor measurement data from smartphones, and present a comprehensive set of de-noising techniques (Task 1. Data Collection and Preprocessing). Second, we design feature extraction methods that carefully balance between prediction accuracy, calculation time, and battery consumption (Task 2. Feature Engineering). Third, we develop new learning models that employ a hierarchical framework with dynamic sensor selection mechanisms, consider each user's personalized travel behavior, and adapt to newly collected data for a given user (Task 3. Learning Models). With carefully designed experiments, we validate the proposed methods and examine their effectiveness in addressing the aforementioned challenges. Finally, although the three learning models are verified through the travel mode detection application. The basic idea behind these models are applicable in many other smartphone sensor based data mining applications such as indoor localization, user activity recognition, biometric information identification, etc.

© The Author(s), under exclusive license to Springer Nature Switzerland AG 2018   51
Y. Yao et al., *Mobile Data Mining*, SpringerBriefs in Computer Science,
https://doi.org/10.1007/978-3-030-02101-6_7

7.2 Discussions

7.2.1 More Combinations of Sensors

Smartphone sensors can be further combined to certain tasks. Take the travel mode detection task as an example. Because the smartphone GPS sensor is quite weak in subways, inaccurate in urban areas, and is energy consuming, we did collect but did not use GPS data for the model training and prediction. However, by analyzing the GPS data on its own, it reveals to us whether the drive is on local or highway. In Android phones, the GPS modular provides the estimated speed [49] as well as the longitude and latitude of the current position. Among all of the differences between driving in freeway and local streets, a simple fact is that the driving speed can drop to 0 for many times when driving on local roads. This is merely happening during freeway driving unless there's a traffic jam, which is rare and not considered in our discussion. Figure 7.1 shows the GPS estimated speed for a vehicle's local and freeway drive, respectively. We can see that in addition to the fact that freeway drive has a higher speed most of the time, the speed of local drive reaches 0 for several times while the freeway drive does not show such aspects. This information does not need a very accurate speed estimation or strong GPS signal, nor does it require a high sampling rate. Therefore, we can use GPS at very low frequency to detect whether the vehicle is on freeway or not.

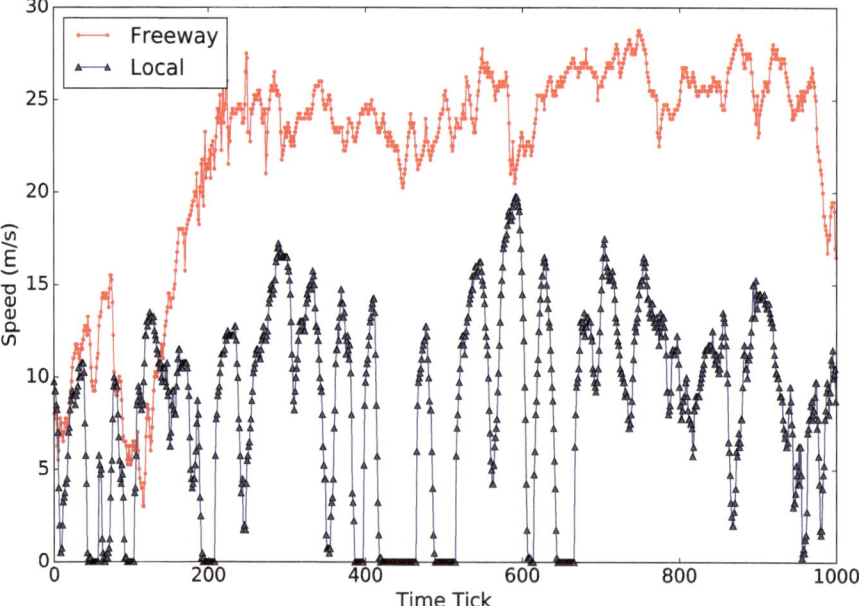

Fig. 7.1 Driving speed at local and freeway

7.2.2   More Usages of Smartphone Sensors

Sensor readings can be affected by the environmental factors, which in turn can help applications if such effect are taken into consideration. For example, we use the barometer data to get the information of ambient pressure, which is an important feature in identifying the different transportation modes. The air pressure also reveals some weather information such as raining, sunny, hot, or cold. For instance, winter days have higher pressure than summer days, the barometer readings recorded in different weather situations show this pattern. Figure 7.2 shows the barometer reading while riding a bike in summer and winter. This could be used as a reference for transportation survey record, as an indication of weather. Further research is needed in order to learn more weather information from the barometer readings.

Similarly, the light sensor which senses the ambient illumination, can also be used. A smartphone provides some default value for different environment conditions such as night without light, cloudy, or sunny day [49]. This can also be a reference in transportation survey. However, one may argue that the sensed ambient light also heavily depends on where the user put the phone: inside a pocket or in hands, and the two ways could result in totally different readings even at the same place. It needs further work and other information to draw the conclusion.

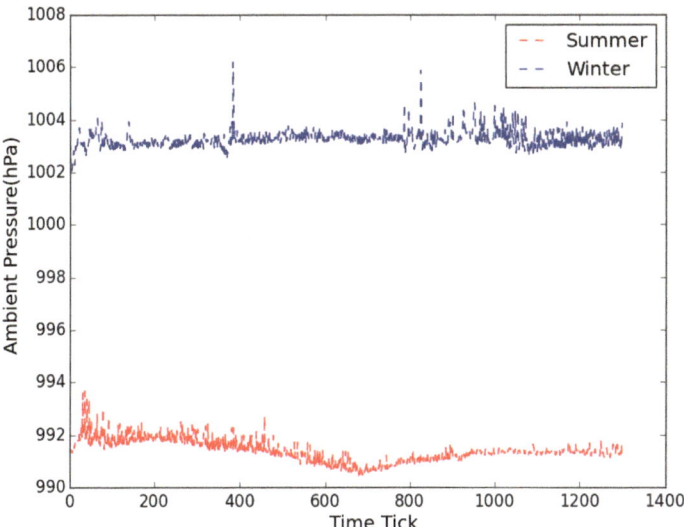

Fig. 7.2 Air pressure at summer and winter on bike

References

1. (2016) URL https://mysleepbot.com/
2. (2016) URL http://stepzapp.com/
3. Abbasi A, Rashidi TH, Maghrebi M, Waller ST (2015) Utilising location based social media in travel survey methods: bringing twitter data into the play. In: Proceedings of the 8th ACM SIGSPATIAL International Workshop on Location-Based Social Networks, ACM, p 1
4. Android (2015) Sensors overview | android developers. http://developer.android.com/guide/topics/sensors/sensors_overview.html, (Visited on 09/02/2015)
5. Android (2016) Sensor | android developers. https://developer.android.com/reference/android/hardware/Sensor.html#TYPE_LIGHT, (Accessed on 03/13/2017)
6. Android (2016) Sensor | android developers. https://developer.android.com/reference/android/hardware/Sensor.html#TYPE_MAGNETIC_FIELD, (Accessed on 03/13/2017)
7. Apple (2016) magneticfield - cmmagnetometerdata | apple developer documentation. https://developer.apple.com/reference/coremotion/cmmagnetometerdata/1616084-magneticfield, (Accessed on 03/13/2017)
8. Apple (2016) Motion events. https://developer.apple.com/library/content/documentation/EventHandling/Conceptual/EventHandlingiPhoneOS/motion_event_basics/motion_event_basics.html, (Accessed on 03/13/2017)
9. Apple (2017) brightness - uiscreen | apple developer documentation. https://developer.apple.com/reference/uikit/uiscreen/1617830-brightness, (Accessed on 03/13/2017)
10. Apple (2017) Iokit.framework - iphone development wiki. http://iphonedevwiki.net/index.php/IOKit.framework, (Accessed on 03/13/2017)
11. Bao L, Intille SS (2004) Activity recognition from user-annotated acceleration data. In: Pervasive computing, Springer, pp 1–17
12. Becker R, Cáceres R, Hanson K, Isaacman S, Loh JM, Martonosi M, Rowland J, Urbanek S, Varshavsky A, Volinsky C (2013) Human mobility characterization from cellular network data. Communications of the ACM 56(1):74–82
13. Benjamin B, Erinc G, Carpin S (2015) Real-time wifi localization of heterogeneous robot teams using an online random forest. Autonomous Robots 39(2):155–167
14. Bottou L (2012) Stochastic gradient descent tricks. In: Neural Networks: Tricks of the Trade, Springer, pp 421–436
15. Center NGD (2014) URL http://www.ngdc.noaa.gov/geomag-web/#igrfwmm
16. Chakravarty T, Ghose A, Bhaumik C, Chowdhury A (2013) Mobidrivescore - a system for mobile sensor based driving analysis: A risk assessment model for improving one's driving. In: Sensing Technology (ICST), 2013 Seventh International Conference on, IEEE, pp 338–344

17. Chen C, Gong H, Lawson C, Bialostozky E (2010) Evaluating the feasibility of a passive travel survey collection in a complex urban environment: Lessons learned from the new york city case study. Transportation Research Part A: Policy and Practice 44(10):830–840

18. Chen PT, Hsieh HP (2012) Personalized mobile advertising: Its key attributes, trends, and social impact. Technological Forecasting and Social Change 79(3):543–557

19. Chon J, Cha H (2011) Lifemap: A smartphone-based context provider for location-based services. IEEE Pervasive Computing (2):58–67

20. De Nunzio G, Wit CC, Moulin P, Di Domenico D (2015) Eco-driving in urban traffic networks using traffic signals information. International Journal of Robust and Nonlinear Control

21. Enck W, Gilbert P, Han S, Tendulkar V, Chun BG, Cox LP, Jung J, McDaniel P, Sheth AN (2014) Taintdroid: an information-flow tracking system for realtime privacy monitoring on smartphones. ACM Transactions on Computer Systems (TOCS) 32(2):5

22. Feng C, Au WSA, Valaee S, Tan Z (2012) Received-signal-strength-based indoor positioning using compressive sensing. Mobile Computing, IEEE Transactions on 11(12):1983–1993

23. Ferris B, Fox D, Lawrence ND (2007) Wifi-slam using gaussian process latent variable models. In: IJCAI, vol 7, pp 2480–2485

24. Google (2009) Coding for battery life. https://dl.google.com/io/2009/pres/W_0300_CodingforLife-BatteryLifeThatIs.pdf, (Accessed on 04/25/2017)

25. Grabowicz PA, Ramasco JJ, Gonçalves B, Eguíluz VM (2014) Entangling mobility and interactions in social media. PLoS One 9(3):e92,196

26. Habib MA, Mohktar MS, Kamaruzzaman SB, Lim KS, Pin TM, Ibrahim F (2014) Smartphone-based solutions for fall detection and prevention: challenges and open issues. Sensors 14(4):7181–7208

27. Huang J, Smola AJ, Gretton A, Borgwardt KM, Scholkopf B (2006) Correcting sample selection bias by unlabeled data. In: NIPS, pp 601–608

28. Huang J, Millman D, Quigley M, Stavens D, Thrun S, Aggarwal A (2011) Efficient, generalized indoor wifi graphslam. In: Robotics and Automation (ICRA), 2011 IEEE International Conference on, IEEE, pp 1038–1043

29. Jahangiri A, Rakha HA (2015) Applying machine learning techniques to transportation mode recognition using mobile phone sensor data. IEEE transactions on intelligent transportation systems 16(5):2406–2417

30. Katevas K, Haddadi H, Tokarchuk L (2016) Sensingkit: Evaluating the sensor power consumption in ios devices. In: Intelligent Environments (IE), 2016 12th International Conference on, IEEE, pp 222–225

31. Khan AM, Lee YK, Lee S, Kim TS (2010) Human activity recognition via an accelerometer-enabled-smartphone using kernel discriminant analysis. In: Future Information Technology (FutureTech), 2010 5th International Conference on, IEEE, pp 1–6

32. Khan AM, Lee YK, Lee SY, Kim TS (2010) A triaxial accelerometer-based physical-activity recognition via augmented-signal features and a hierarchical recognizer. Information Technology in Biomedicine, IEEE Transactions on 14(5):1166–1172

33. Kwapisz JR, Weiss GM, Moore SA (2011) Activity recognition using cell phone accelerometers. ACM SigKDD Explorations Newsletter 12(2):74–82

34. Ladd AM, Bekris KE, Rudys A, Kavraki LE, Wallach DS (2005) Robotics-based location sensing using wireless ethernet. Wireless Networks 11(1–2):189–204

35. Lathia N, Capra L (2011) Mining mobility data to minimise travellers' spending on public transport. In: Proceedings of the 17th ACM SIGKDD international conference on Knowledge discovery and data mining, ACM, pp 1181–1189

36. Le LT, Eliassi-Rad T, Provost F, Moores L (2013) Hyperlocal: inferring location of ip addresses in real-time bid requests for mobile ads. In: Proceedings of the 6th ACM SIGSPATIAL International Workshop on Location-Based Social Networks, ACM, pp 24–33

37. LiKamWa R, Liu Y, Lane ND, Zhong L (2011) Can your smartphone infer your mood. In: PhoneSense workshop, pp 1–5

38. Liu K, Liu X, Li X (2013) Guoguo: Enabling fine-grained indoor localization via smartphone. In: Proceeding of the 11th annual international conference on Mobile systems, applications, and services, ACM, pp 235–248

39. Maghdid HS, Lami IA, Ghafoor KZ, Lloret J (2016) Seamless outdoors-indoors localization solutions on smartphones: Implementation and challenges. ACM Computing Surveys (CSUR) 48(4):53

40. Manzoni V, Maniloff D, Kloeckl K, Ratti C (2010) Transportation mode identification and real-time co2 emission estimation using smartphones. SENSEable City Lab, Massachusetts Institute of Technology, nd

41. Martín H, Bernardos AM, Iglesias J, Casar JR (2013) Activity logging using lightweight classification techniques in mobile devices. Personal and ubiquitous computing 17(4):675–695

42. Muralidharan K, Khan AJ, Misra A, Balan RK, Agarwal S (2014) Barometric phone sensors: more hype than hope! In: Proceedings of the 15th Workshop on Mobile Computing Systems and Applications, ACM, p 12

43. Papadimitriou S, Eliassi-Rad T (2015) Mining mobility data. In: Proceedings of the 24th International Conference on World Wide Web Companion, International World Wide Web Conferences Steering Committee, pp 1541–1542

44. Physikalisch-Technische Bundesanstalt G Braunschweig (2014) URL http://www.ptb.de/cartoweb3/SISproject.php

45. Provost FJ, Eliassi-Rad T, Moores LS (2015) Methods, systems, and media for determining location information from real-time bid requests. US Patent 9,014,717

46. Raento M, Oulasvirta A, Petit R, Toivonen H (2005) Contextphone: A prototyping platform for context-aware mobile applications. Pervasive Computing, IEEE 4(2):51–59

47. Ratti C, Sobolevsky S, Calabrese F, Andris C, Reades J, Martino M, Claxton R, Strogatz SH (2010) Redrawing the map of great britain from a network of human interactions. PloS one 5(12):e14,248

48. Ravi N, Dandekar N, Mysore P, Littman ML (2005) Activity recognition from accelerometer data. In: Proceedings of the national conference on artificial intelligence, Menlo Park, CA; Cambridge, MA; London; AAAI Press; MIT Press; 1999, vol 20, p 1541

49. Reference AD (2014) URL http://developer.android.com/reference/android/hardware/SensorEvent.html#values

50. Rissel C, Curac N, Greenaway M, Bauman A (2012) Physical activity associated with public transport use: a review and modelling of potential benefits. International journal of environmental research and public health 9(7):2454–2478

51. Rossi M, Seiter J, Amft O, Buchmeier S, Tröster G (2013) Roomsense: an indoor positioning system for smartphones using active sound probing. In: Proceedings of the 4th Augmented Human International Conference, ACM, pp 89–95

52. Rumelhart DE, Hinton GE, Williams RJ (1985) Learning internal representations by error propagation. Tech. rep., DTIC Document

53. Saad D (1998) Online algorithms and stochastic approximations. Online Learning 5

54. Sadilek A, Kautz H, Bigham JP (2012) Finding your friends and following them to where you are. In: Proceedings of the fifth ACM international conference on Web search and data mining, ACM, pp 723–732

55. Sankaran K, Zhu M, Guo XF, Ananda AL, Chan MC, Peh LS (2014) Using mobile phone barometer for low-power transportation context detection. In: Proceedings of the 12th ACM Conference on Embedded Network Sensor Systems, ACM, pp 191–205

56. Schölkopf B, Smola AJ (2002) Learning with kernels: support vector machines, regularization, optimization, and beyond. MIT press

57. Shaheen S, Guzman S, Zhang H (2010) Bikesharing in europe, the americas, and asia: past, present, and future. Transportation Research Record: Journal of the Transportation Research Board (2143):159–167

58. Shalev-Shwartz S, Singer Y, Srebro N, Cotter A (2011) Pegasos: Primal estimated sub-gradient solver for svm. Mathematical programming 127(1):3–30

59. Toole JL, Herrera-Yaqüe C, Schneider CM, González MC (2015) Coupling human mobility and social ties. Journal of The Royal Society Interface 12(105):20141,128

60. Varnali K, Toker A (2010) Mobile marketing research: The-state-of-the-art. International journal of information management 30(2):144–151

61. Wu C, Yang Z, Liu Y (2015) Smartphones based crowdsourcing for indoor localization. Mobile Computing, IEEE Transactions on 14(2):444–457

62. Xie H, Gu T, Tao X, Ye H, Lv J (2014) Maloc: A practical magnetic fingerprinting approach to indoor localization using smartphones. In: Proceedings of the 2014 ACM International Joint Conference on Pervasive and Ubiquitous Computing, ACM, pp 243–253

63. Xing Su HT Hernan Caceres, He Q (2015) Travel mode identification with smartphones. Transportation Research Board Annual Meeting

64. Xu Z, Bai K, Zhu S (2012) Taplogger: Inferring user inputs on smartphone touchscreens using on-board motion sensors. In: Proceedings of the fifth ACM conference on Security and Privacy in Wireless and Mobile Networks, ACM, pp 113–124

65. Ye H, Gu T, Zhu X, Xu J, Tao X, Lu J, Jin N (2012) Ftrack: Infrastructure-free floor localization via mobile phone sensing. In: Pervasive Computing and Communications (PerCom), 2012 IEEE International Conference on, IEEE, pp 2–10

66. Ye H, Gu T, Tao X, Lu J (2014) Sbc: Scalable smartphone barometer calibration through crowdsourcing. In: Proceedings of the 11th International Conference on Mobile and Ubiquitous Systems: Computing, Networking and Services, ICST (Institute for Computer Sciences, Social-Informatics and Telecommunications Engineering), pp 60–69

67. Yuan J, Zheng Y, Zhang C, Xie W, Xie X, Sun G, Huang Y (2010) T-drive: driving directions based on taxi trajectories. In: Proceedings of the 18th SIGSPATIAL International conference on advances in geographic information systems, ACM, pp 99–108

68. Yuan J, Zheng Y, Xie X (2012) Discovering regions of different functions in a city using human mobility and pois. In: Proceedings of the 18th ACM SIGKDD international conference on Knowledge discovery and data mining, ACM, pp 186–194

69. Yuan M, Lin Y (2006) Model selection and estimation in regression with grouped variables. Journal of the Royal Statistical Society: Series B (Statistical Methodology) 68(1):49–67

70. Zheng Y, Liu Y, Yuan J, Xie X (2011) Urban computing with taxicabs. In: Proceedings of the 13th international conference on Ubiquitous computing, ACM, pp 89–98

71. Zheng Y, Capra L, Wolfson O, Yang H (2014) Urban computing: concepts, methodologies, and applications. ACM Transactions on Intelligent Systems and Technology (TIST) 5(3):38